CHEMICAL BONDING

CHEMICAL BONDING

SECOND EDITION

AUDREY L. COMPANION
Department of Chemistry
University of Kentucky

McGRAW-HILL BOOK COMPANY

New York St. Louis San Francisco Auckland Bogotá Düsseldorf
Johannesburg London Madrid Mexico Montreal New Delhi
Panama Paris São Paulo Singapore Sydney Tokyo Toronto

CHEMICAL BONDING

234567890 MUMU 7832109

This book was set in Times Roman by A Graphic Method Inc.
The editors were Donald C. Jackson and Stephen Wagley;
the designer was Nicholas Krenitsky;
the production supervisor was Leroy A. Young.
The Murray Printing Company was printer and binder.

Library of Congress Cataloging in Publication Data

Companion, Audrey L
Chemical bonding.

Includes bibliographies and index.
1. Chemical bonds. I. Title.
QD461.C63 1979 541'.224 78-17081
ISBN 0-07-012383-7
ISBN 0-07-012379-9 (pbk.)

CONTENTS

PREFACE

When the first edition of *Chemical Bonding* appeared in 1964, its major purpose was clear: to correct a deficiency in orbital theory in most general chemistry textbooks. Since most textbooks today are more than adequate in treatment of atomic and molecular structure, obviously a remedial supplement is no longer needed. The audience for the book has for some years now included advanced high school students who wish more detail, freshman college students seeking a "how-to-do-it" approach, and upper-level students wanting a broad qualitative overview of bonding theory before plunging into quantum-mechanical treatments.

In this edition there is virtually no change in the level of presentation of the material. After many years of teaching bonding theory to freshman students I believe that explicit discussions of the characteristics of wave functions, flooded-planet models, signs of *LCAOMO*s, etc., are not understood by the average student. An instructor with better-than-average students can easily expand such topics in lecture.

The first two chapters, covering the experimental background and history of quantum theory, are purposely brief and may be omitted. Large portions of Chapter 3 on atoms and Chapter 4 on molecules have been rewritten, and most diagrams are new. In Chapter 4 I have retained the mixed *VB–MO* approach of the first edition primarily because it works well with students. Chapter 5 has been expanded, particularly in sections on metals and semiconductors, and a new section on silicates and glasses has been added to demonstrate materials science applications of the theory. More advanced concepts on transition-metal chemistry, such as the Jahn-Teller effect, have been removed from Chapter 6 in this edition.

Other general changes include: incorporation of certain topics (such as dative bonding), previously developed in the text, into problems at the chapter ends; an increase in the number of problems and revision of most of the old ones; use of SI units throughout; and an increase in the number of journal references.

Through the years the suggestions and criticisms offered by friends at many universities have led to what I hope now is a better book. In particular I wish to thank my former colleague Kenneth Schug for his help both in the early years and in the development of the materials science ideas included in this edition, my present colleague Paul Corio for his witch hunt on Pennsylvania Duchisms and for other comments, and Mary and Martin Kilpatrick for their encouragement over many years. I am grateful to the following reviewers for their helpful comments upon reading the manuscript: Professor David L. Adams, North Shore Community College; Professor Frank O. Ellison, University of Pittsburgh; Professor Gil Haight, University of Illinois at Urbana-Champaign; Professor Robert M. Kren, University of Michigan-Flint; Dr. Conrad Stanitski, Randolph-Macon College; and Professor Duane D. Swank, Pacific Lutheran University. All errors remaining are my own.

Audrey L. Companion

CHEMICAL BONDING

THE BEGINNINGS OF QUANTUM THEORY

1

INTRODUCTION

1-1 Frustration and rebellion, though of a quiet sort, abounded in science in the early twentieth century, particularly among those concerned with the nature of matter and energy. Many new ideas were born, and many promptly buried; many old and established laws of physics were shaken. Out of this chaotic period emerged the modern theory of the structure of atoms, molecules, and solids, a theory virtually unchallenged today. A rigorous discussion of its basis requires at least a sound understanding of calculus, a tool usually not at the fingertips of beginners in chemistry. Yet even without the underlying mathematics we can describe quite well the nature of atoms and molecules, since the physical theory is rich in pictures and rules which are usually easily accepted. These we shall lean upon heavily in this text. Frequently, though, we shall encounter concepts which seem unpalatable, for they defy the rules governing events occurring in everyday life. Yet accept them we must, for they are supported by unequivocal experimental evidence. One of these unusual concepts involves the dual life led by the phenomenon called light, with which we begin.

THE NATURE OF LIGHT

1-2 Largely because of the impact of the creative genius of Sir Isaac Newton (1642–1727), who advocated a corpuscular (particle-like) model, the wave theory of light propagation was not really accepted until about 1850, despite the accumulation of experimental evidence supporting it. At that time the experiments finally overwhelmed the particle model, and until the turn of the century the wave theory was undisputed. Many scientists felt that the corpuscular model had been properly and permanently put to rest.

Even now it is believed that light is *propagated through space* in the form of a wave motion, somewhat like ripples on a pond at the drop of a pebble. Figure 1-1 illustrates a representation of such a wave. The distance A is the maximum *amplitude* of the disturbance; the distance from crest to crest (or valley to valley) is its *wavelength* λ (Greek lambda), a distance quite large in a pond but very small when the wave motion describes light. For example, for visible light λ ranges from 400 to 700 nm.

Light travels through space with a velocity c of approximately 3×10^8 m/s; i.e., the peaks and valleys of Fig. 1-1 move in the direction of the light beam with a velocity c. In 1 sec a stationary microcosmic observer of a light beam of wavelength λ would count c/λ peaks passing by or would observe a frequency of peaks ν (Greek nu) of c/λ cycles per second associated with the wave. Thus, for *light*, wavelength and

FIGURE 1-1
Representation of a light wave.

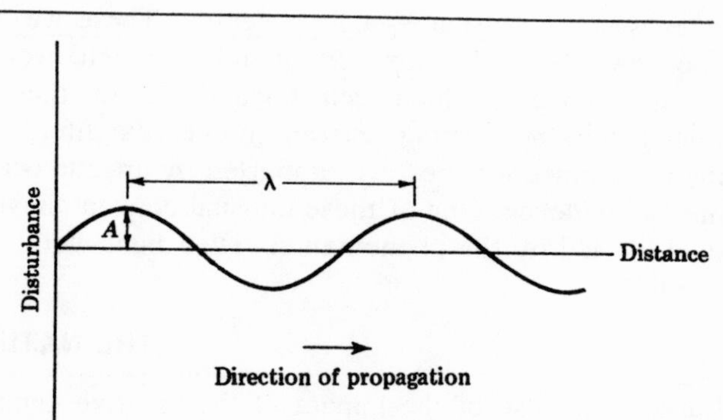

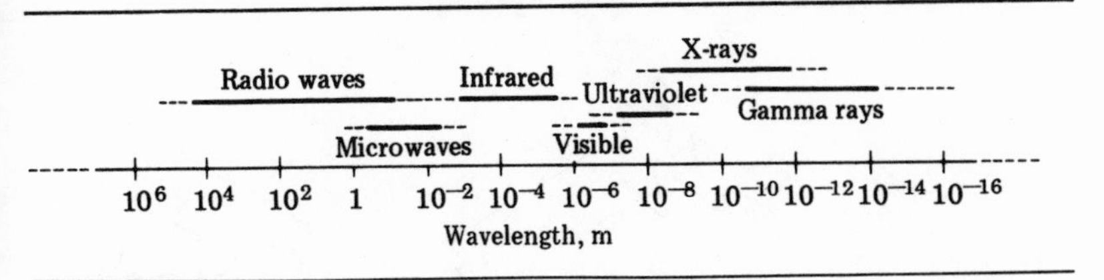

FIGURE 1-2
The electromagnetic spectrum.

frequency, both convenient descriptions of the wave property, are related by the equation

$$\lambda \nu = c$$

That which we call visible light is but a very narrow part of a large group of radiation types comprising the *electromagnetic spectrum* (Fig. 1-2), which includes very-long-wavelength radio waves and very-short-wavelength gamma rays. All these radiation types are the same phenomenon; the basis for the classification shown in Fig. 1-2 is largely the experimental means of detection or generation.

The human eye is tuned to interpret only radiation of wavelength 400 to 700 nm (where 1 nm $= 10^{-9}$ m), and these numbers are thus the limits of the visible portion of the electromagnetic spectrum.

There are many experimental "proofs" of the wave nature of light. One, which will be of particular use to us later, involves the reflection or diffraction of x-rays by orderly stacks of atoms in a crystal. Figure 1-3 shows in cross section two x-rays of wavelength λ impinging at an angle θ on the surface of a crystal in which atoms are arrayed in planes separated by a distance d. Both rays are associated with waves traveling in phase, so that their amplitudes maximize and minimize together and they reinforce one another (point A), at least up to the line BE, after which the rays undergo reflections from different planes. Now unless the distance BCD is equal to λ or to some integral multiple n of λ, the two emerging rays will be out of phase and may cancel one another. Through simple geome-

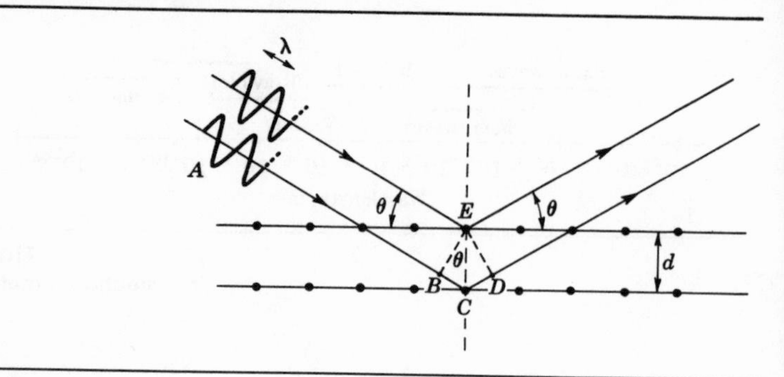

FIGURE 1-3
Diffraction of x-rays by a crystal lattice.

try *BCD* is $2d \sin \theta$, so that the condition for getting strong reflections from the crystal is

$$n\lambda = 2d \sin \theta \quad \text{where } n = 0, 1, 2, 3, \ldots$$

the Bragg diffraction law. Experimentally, if the angle of incidence of x-rays on a crystal surface is varied, strong reflections are observed at just the angles predicted, with blackness or grayness in between. Such an experiment could not be explained by a particle model of light.

However, in 1900 the German scientist Max Planck reintroduced the concept of corpuscular light while presenting a theory explaining blackbody radiation. If the radiation emerging from a pinhole in a very hot closed furnace, a "black body," is passed through a prism and a graph of amount of energy emitted versus wavelength is constructed for a given temperature, a curve like one of those in Fig. 1-4 is obtained. Explanations of these curves based on all the rules and regulations of physics known at the time (classical physics) failed to explain the shapes and temperature dependence of the curves.

Planck attacked this problem by searching for a formula connecting radiant energy, temperature, and wavelength, guessing different functions and adjusting them with numerical constants until he found the correct *empirical* relationship between the variables. Armed with this, he then searched for a hypothetical "model" for the furnace system from which he could derive theoretically his empirical formula. Success came rapidly when he compared the atoms constituting the walls of the furnace to a large assembly of oscillators of all vibrational frequencies absorbing and emitting energy. One of the assumptions of his analysis was startling: the oscillators could change their energy by absorbing or emitting only spurts or bundles of energy, which he called *quanta*. Furthermore, a quantum of energy was related to the oscillator frequency ν by the equation $E = h\nu$, where h is a proportionality constant, Planck's constant. When the absorption and emission probabilities of the group of oscillators were counted, radiant energy distributions like those of Fig. 1-4 resulted. Such distributions could not be obtained without these assumptions.

Until this time it was believed that a vibrating body could change its energy by an arbitrary amount (say 0.111 $h\nu$, 0.697 $h\nu$, etc.). Quite possibly Planck's ideas would not have been accepted had it not been for Einstein's use of the quantum concept 5 years later in the explanation of the *photoelectric effect*.

FIGURE 1-4
Blackbody radiation.

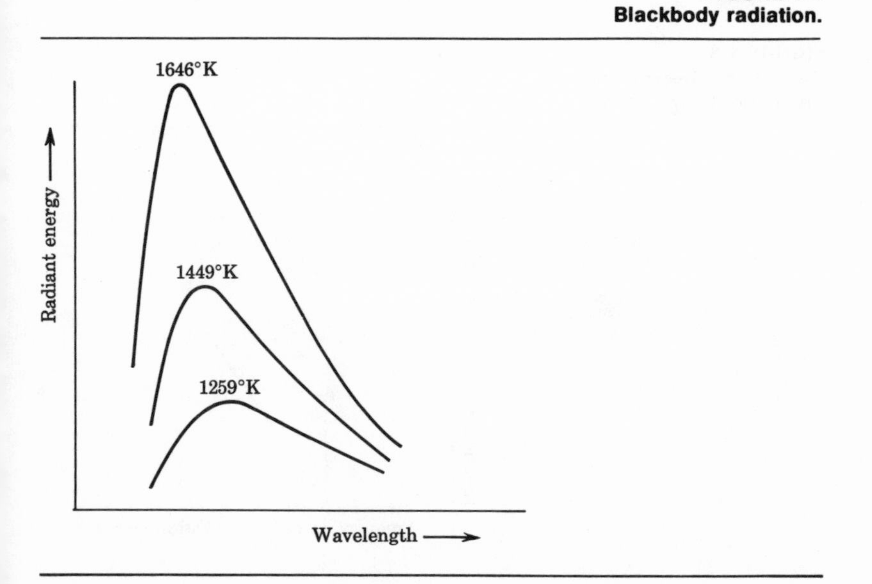

When a clean surface of an alkali metal was irradiated with a light beam, electrons were ejected from the surface. Studies of the kinetic energy ($\frac{1}{2}mv^2$) of the ejected electrons as a function of light frequency ν indicated that below a critical frequency ν_0, characteristic of the metal, no electrons were ejected. For frequencies above ν_0, the kinetic energy of the electrons increased with ν linearly (see Fig. 1-5). These data defied explanation in terms of the wave model. Einstein explained the phenomenon quite simply in terms of collisions between electrons in the metal and bundles of light of energy $E = h\nu$ which he called *photons*. When a bombarding photon hits the surface it collides with and transfers its energy to an electron. The electron expends some of this energy in work (W) to escape the metal, retaining the rest as kinetic energy. Einstein's equation for the process was:

$$\underbrace{h\nu}_{\text{photon}} = \underbrace{W + \tfrac{1}{2}mv^2}_{\text{electron}}$$

For photon energies below the constant work term $W = h\nu_0$, no electrons can escape the metal. Photon energies in excess of W appear as kinetic energy of the electron ejected after the collision. A collision process is a corpuscular phenomenon!

FIGURE 1-5
Data from observations of the photoelectric effect.

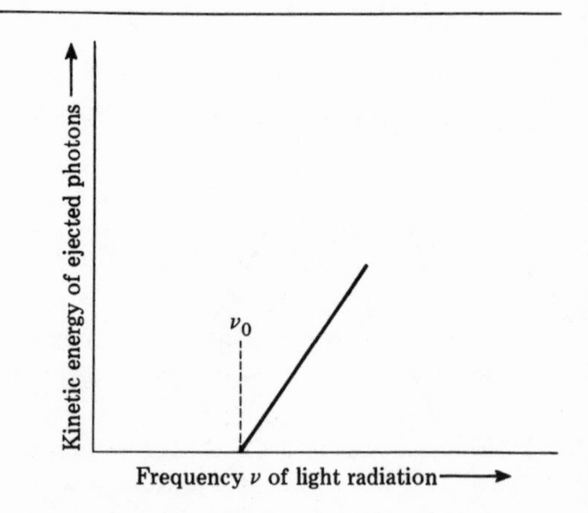

Both models, corpuscular and wave, seem here to stay. Light, as it is propagated through space, is distinctly wavelike and, when it interacts with matter in energy-exchange processes, sometimes assumes a corpuscular nature.

We shall meet the important proportionality constant h again in the Bohr model of the atom (Chap. 2) and in the Schrödinger equation (Chap. 3), on which we base the modern theory of atoms and molecules and solids. h is now regarded as a fundamental constant of nature, particularly appropriate to systems of atomic size. In SI units h is 0.6626×10^{-33} J/s.

With the relation $E = h\nu$, we can now discuss the electromagnetic spectrum in terms of energy. Since $\nu\lambda = c$, $E = hc/\lambda$, and since both h and c are constants, the energy of a photon is inversely proportional to its wavelength. Radio waves are of low energy, gamma rays of high energy; near and within the visible range, infrared (IR) and red are of lower energy, violet and ultraviolet (UV) of higher energy.

The spectrum emitted by a furnace (Fig. 1-4) is a *continuum*, a display of *all* wavelengths, with no detectable gaps. Many heated solids, for example, a tungsten filament in an electric bulb, emit such a continuum. This smear of all wavelengths appears as white light to an observer.

THE DUAL NATURE OF MATTER

1-3

The notion of the electron as a particle is commonly accepted and has been for a long time now. Its wavelike character is less easily conceived.

Basing his arguments on the symmetry of nature, the French physicist de Broglie postulated that, if light has both particlelike and wavelike character, a similar duality must exist for matter, and he proceeded to show that a definite wavelength could be associated with the movement of bodies of matter.

Previously Einstein had proved theoretically that mass and energy were interconvertible quantities (a theory later proved experimentally by nuclear physicists and chemists) and had shown that associated with a photon of energy E was an equivalent mass of E/c^2. The momentum p (mass $\times$ velocity) of a photon (a corpuscular property) is thus related to its wavelength (obviously a wave property) by

$$p = \text{mass} \times \text{velocity} = \frac{E}{c^2}c = \frac{h\nu}{c} = \frac{h}{\lambda}$$

i.e., momentum and wavelength are inversely proportional through Planck's constant.

De Broglie postulated that with particles of matter of mass m and velocity v was associated a wavelength λ given by

$$\lambda = \frac{h}{p} = \frac{h}{mv}$$

a relationship soon proved experimentally for electrons. A pair of scientists, Davisson and Germer, verified the de Broglie postulate by studying a beam of electrons accelerated to a high, well-defined kinetic energy (and thus with well-known momentum and λ). When directed at the surface of a nickel crystal, the electron beam was reflected in much the same way as an x-ray beam; reinforcement and interference of the wavelike electron rays led to detectable electron scattering at angles predicted by an equation similar to that of Bragg.

Thus physicists resigned themselves to the fact that electrons too have Jekyll-and-Hyde characters, sometimes behaving like particles and at other times like waves. In the ensuing chapters we shall use both concepts of the electron: a small, compact, negatively charged particle and a smeared-out, wavelike, negatively charged cloud. The reader is forewarned to accept both.

EXERCISES

1 Photons of (red, violet) light are of higher energy and photons of (red, violet) light have the longer wavelength.

2 Bragg's diffraction law requires a (wave, corpuscular) interpretation of light.

3 Explanation of the photoelectric effect requires a (wave, corpuscular) interpretation of light.

4 A particle will have associated with it a longer wavelength when it is moving at (10^3 m/s, 10^6 m/s).

5 The mass of a proton is 1.67×10^{-24} g while that of an electron is 9.11×10^{-28} g. In which situation would the particle have the longest wavelength?

(*a*) A proton moving with velocity 10^6 m/s
(*b*) A proton moving with velocity 10^4 m/s
(*c*) An electron moving with velocity 10^6 m/s
(*d*) An electron moving with velocity 10^4 m/s

6 Calculate the wavelength associated with an electron moving at 1.5×10^6 m/s.

7 Calculate the energy associated with a photon of light of wavelength 400 nm.

8 A light beam irradiates simultaneously the surfaces of two metals *A* and *B*. At wavelength λ_1, electrons are ejected only from metal *A*. At wavelength λ_2, metals *A* and *B* eject equal numbers of photons. Consequently,

(*a*) The wavelength λ_1 is (shorter, longer) than λ_2.
(*b*) Electrons need more energy to escape metal (*A*, *B*).
(*c*) Under λ_2 irradiation the kinetic energy of electrons emitted from *A* is (greater, less) than the kinetic energy of electrons from *B*.
(*d*) Electrons emitted from *A* have the greater kinetic energy when produced by (λ_1, λ_2) light.

REFERENCES FOR FURTHER READING

1 Barrow, G. M., "The Structure of Molecules: An Introduction to Molecular Spectroscopy," chap. 1, W. A. Benjamin, Inc., New York, 1963.

2 Bragg, Sir Lawrence, The Start of X-ray Analysis, *Chemistry*, **40,** 8, December (1967).

3 Davis, Jeff C., Jr., Introduction to Spectroscopy: Part III. Light and the Electromagnetic Spectrum, *Chemistry*, **48,** 19, May (1975).

THE BOHR ATOM

2

INTRODUCTION

2-1

By 1913 the nature of the building blocks of the atom was well known. In a series of independent experiments J. J. Thomson and R. A. Millikan had determined that the electron was a small, negatively charged particle of mass about 0.9×10^{-30} kg and charge 0.16×10^{-18} C. By bombarding a thin gold foil with a beam of α particles, or nuclei of helium atoms, Rutherford had shown that most of the mass of the atom was concentrated in a small, positively charged body surrounded by largely empty space and had postulated that the very small electrons were in some way located in this space, making the atom neutral in charge. Now obviously the electrons could not be at rest, for electrostatic attraction would quickly suck them into the nucleus. Rutherford proposed that electrons were whirling at very high velocities in circular paths around the nucleus, so that the outward pull associated with such a motion would counteract the nuclear pull. (Compare the outward pull on a bucket of water, counteracting gravitational pull, as it is swung in a vertical circle.) Similar, very successful laws were known to govern the motion of planets around the sun. Unfortunately for the theory, electrons differ from planets in that they are charged particles, and, according to other successful laws of classical physics, an accelerated charged particle will continuously radiate and lose its energy. The orbiting electron was thus doomed to spiral into the nucleus. All attempts to build a model of the atom consistent with the experimental facts and with the laws of physics known and accepted at that time led to disaster. As Gamow (1) said, "it looked for a while as though either the physicists or physics itself had become completely insane."

It remained for the young physicist Niels Bohr to suggest a cure, but at the price of refuting some of the old and well-established laws of physics. Bohr's revolutionary theory of the atom was prompted by his interest in certain experimental facts about the hydrogen atom, which we shall now examine.

OCCURRENCE OF LINE SPECTRA

2-2 In Chap. 1, the radiation emitted by a heated solid body was described as a continuous spectrum of all wavelengths and energies. Figure 2-1 illustrates the *line spectra* obtained when one passes through a prism or grating the light emitted by strongly heated hydrogen atoms. Not all energies of light are observed, but instead certain very discrete ones, which appear in the spectrum as sharp lines separated by blackness. The energies of the lines are characteristic of the identity of the atoms heated, and in fact the more prominent ones in the visible spectrum are the basis for the familiar flame tests used in qualitative analysis. The line spectra in Fig. 2-1 belong to the hydrogen atom, and one can observe in the diagram several clusters or series of lines.

By a purely empirical approach J. J. Balmer found that the cluster of lines occurring in the visible and near ultraviolet had

FIGURE 2-1
Line spectra characteristic of the hydrogen atom.

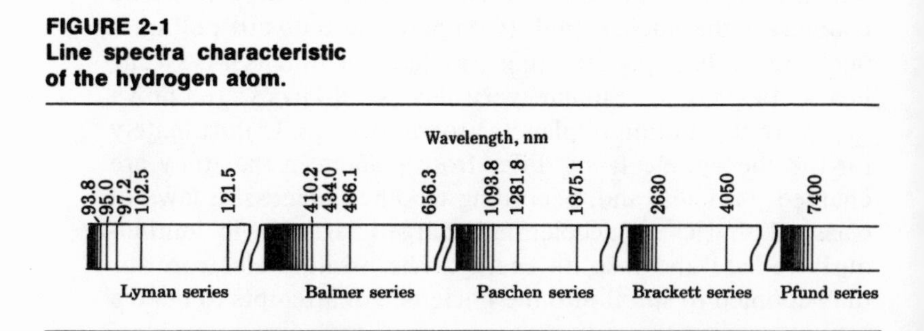

wavelengths which were interrelated by an equation of the form:

$$\frac{1}{\lambda} = \mathbf{R}\left[\left(\frac{1}{2}\right)^2 - \left(\frac{1}{n}\right)^2\right]$$

where n is any integer greater than 2, and **R** is a constant, known as the Rydberg constant, whose value is $1.09737 \times 10^{-2}\ \text{nm}^{-1}$. J. J. Rydberg and others showed that the wavelengths of all observed hydrogen clusters could be accounted for by the more general expression:

$$\frac{1}{\lambda} = \mathbf{R}\left[\left(\frac{1}{n_L}\right)^2 - \left(\frac{1}{n_U}\right)^2\right]$$

where n_U is an integer greater than the integer n_L. n_L takes on the following values for the different clusters: $n_L = 1$, Lyman series; $n_L = 2$, Balmer series; $n_L = 3$, Paschen series; $n_L = 4$, Brackett series; $n_L = 5$, Pfund series. Bohr intended to formulate a theoretical model for the hydrogen atom which would account for these rules.

STRUCTURE OF THE HYDROGEN ATOM

2-3

Bohr accepted Rutherford's notion of electron motion in circular orbits, but he rejected the classical law that accelerated charged bodies radiate energy and instead arbitrarily assumed (1) that the electron while in a particular orbit had a well-defined characteristic energy which *could not change* while it was in that orbit and (2) that only *certain discrete energies were allowed* for the electron. With these assumptions, he was able to show that the energies available to the electron in hydrogen are given by

$$E_n = \frac{-2\pi^2 me^4}{h^2 n^2} = \frac{-K}{n^2} \qquad \text{where } K = \frac{2\pi^2 me^4}{h^2} = 2.18 \times 10^{-18}\ \text{J}$$

Here e is the magnitude of the charge on the electron, m is the electron mass, h is Planck's constant, and n is an integer, called the *principal quantum number*, which can take on the values 1, 2, 3, 4, . . . , each value of n defining a new energy for the electron. Associated with each energy is a circular orbit of well-defined radius around the nucleus given by

$$r_n = \frac{n^2h^2}{4\pi^2me^2} = n^2r_1$$

$$\text{where } r_1 = \frac{h^2}{4\pi^2me^2} = 0.05292 \text{ nm or } 0.5292 \text{ Å}$$

We can see that, as n approaches infinity, the nucleus and the electron are separated by an infinite distance, and consequently their attractive interaction energy must be zero. E_n in this limit is zero. As the orbits get closer to the nucleus (i.e., as n gets smaller), E_n becomes larger in *absolute value*, yet more and more negative. We shall identify the magnitude of E_n with the attractive energy holding the atom together (with the *stability* of the atom); yet because of the sign of E_n (which is determined by arbitrary convention), the most negative energy represents the most stable system with respect to the infinitely separated electron and nucleus. On an energy-level diagram such as Fig. 2-2, increasing energy (toward less negative values) represents decreasing stability. Within this convention the state of lowest energy for the hydrogen atom is that for which $n = 1$. We name this the *ground state* of the system. All other allowed energies shown in Fig. 2-2 represent less stable, *excited* states. No energies between those indicated in Fig. 2-2 are permitted for the electron in the hydrogen atom.

Figure 2-3 shows the variation in orbit radii with n. When the electron is in its ground state ($n = 1$), it is in the orbit closest to the nucleus. According to Bohr theory it can go no closer.

The line spectra of the H atom can easily be interpreted in terms of the energy-level diagram given in Fig. 2-2, if it is assumed that the electron, excited to an upper, less stable energy level E_U by a flame or spark, falls to a more stable, lower state E_L, and in the process emits a photon or light quantum of energy $E_U - E_L$. Simultaneously it moves from a larger orbit with quantum number n_U to a smaller orbit characterized by n_L. Thus the photon energy would be

$$h\nu = E_U - E_L = \frac{2\pi^2me^4}{h^2}\left[\left(\frac{1}{n_L}\right)^2 - \left(\frac{1}{n_U}\right)^2\right]$$

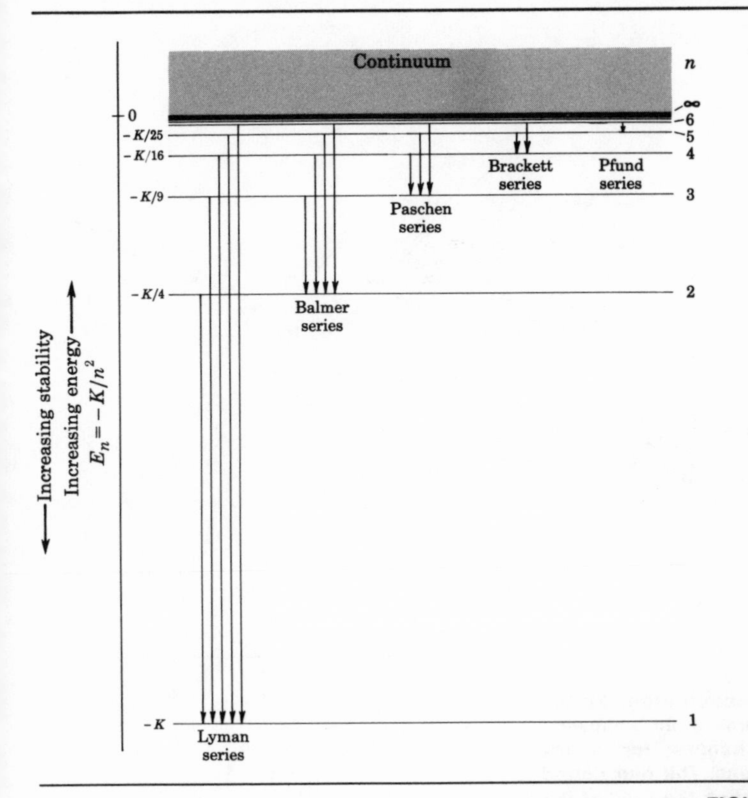

FIGURE 2-2
Energy levels in the hydrogen atom and their relation to spectral series. $K = 2\pi^2 me^4/h^2 = 2.18 \times 10^{-18}$ J.

and accordingly we should have (since $\nu = c/\lambda$)

$$\frac{1}{\lambda} = \frac{2\pi^2 me^4}{h^3 c}\left[\left(\frac{1}{n_L}\right)^2 - \left(\frac{1}{n_U}\right)^2\right]$$

an equation compatible with the Rydberg equation provided that

$$\mathbf{R} = \frac{2\pi^2 me^4}{ch^3}$$

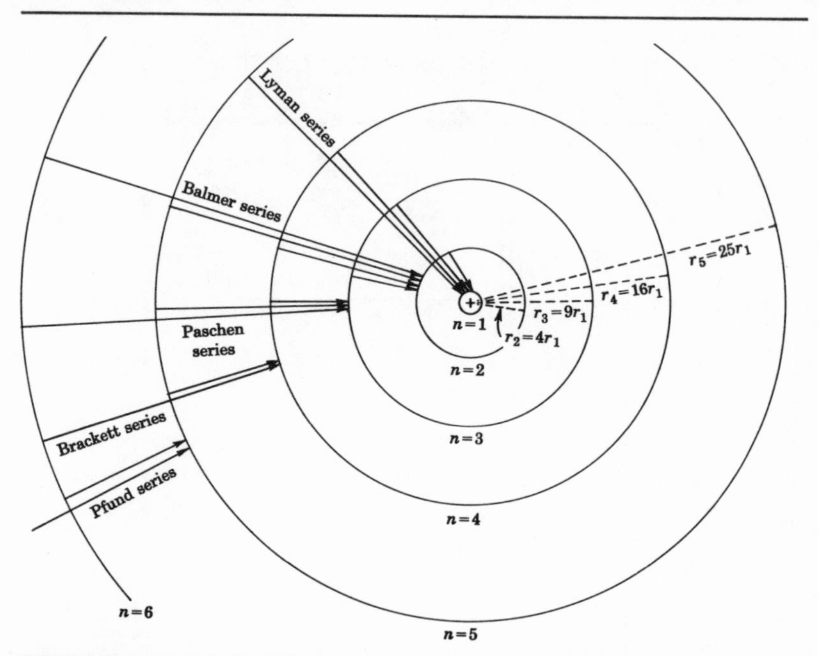

FIGURE 2-3
Bohr electron orbits for the hydrogen atom, showing orbit changes for series transitions. The four dotted lines define the radii of the orbits in terms of the quantum number *n* and the smallest possible orbit r_1. $r_n = n^2 r_1$ where $r_1 = h^2/4\pi^2 me^2 = 0.05292$ nm.

R computed from separately determined values of these constants is in excellent agreement with the empirical value. The identification of the Lyman, Balmer, Paschen, Brackett, and Pfund series in terms of the quantized energy levels provided by Bohr is illustrated in Fig. 2-2.

OTHER ATOMS

2-4 For other *one-electron systems*, such as the He^+ and Li^{2+} ions, the Bohr model is equally good provided that we include the effect of the changed nuclear charge Z in the following way:

$$E_n = Z^2 E_n^{\mathrm{H}} \qquad r_n = \frac{r_n^{\mathrm{H}}}{Z}$$

where E_n^{H} and r_n^{H} are the expressions previously stated for the hydrogen atom. For larger nuclear charges the stabilization energy of the electron is greater and the orbits are closer to the nucleus.

Although remarkably successful for one-electron systems, Bohr theory when applied to many-electron (more than 1) atoms broke down almost completely, and it soon became apparent that more and more arbitrary postulates had to be introduced to bring the model into accord with experimental facts. We shall not pursue these here, for after about 10 years the Bohr model was virtually abandoned in favor of the wave-mechanical theory in prominent use today.

Before closing the story of Bohr, we should pause to reflect on the impact of his work. Though his assumptions of quantization were arbitrary and his results effective for only very simple systems, his courage in abandoning classical laws led (along with work of Einstein and Planck) to the now accepted philosophy that not all physical laws governing the macrocosm of ping-pong balls, people, and planets are applicable to the microcosm of electrons, photons, and particles of similar size. His place as one of the pioneers of the old quantum theory is well deserved. As we shall see, the new quantum theory negated none of his significant results, but rather confirmed the genius of his intuitive assumptions.

EXERCISES

1 Bohr's expression for the binding energy of the electron on the hydrogen atom is $E_n = -K/n^2$, where $K = 2.18 \times 10^{-18}$ J. The corresponding orbit radii are $r_n = n^2 r_1$, where $r_1 = 0.05292$ nm.

(*a*) In multiples of K, what is the energy of the electron in an orbit of radius 4 r_1?

(*b*) In multiples of K, what is the energy of the electron in an orbit of radius 25 r_1?

(*c*) Calculate the energy in joules of the electron when in an orbit of radius 0.4763 nm.

(*d*) The electron is more tightly bound to the nucleus in (case *a*, case *b*).

(*e*) What is the distance between electron and nucleus when $E_n = 0$?

2 The Rydberg relation between all hydrogen atom spectral lines is

$$\frac{1}{\lambda} = \mathbf{R}\left(\frac{1}{n_L^2} - \frac{1}{n_U^2}\right)$$

The shortest wavelength of a photon that can be emitted when an electron jumps from the $n = 4$ state is

$1/\mathbf{R}$ $\quad 16/15\mathbf{R}$ $\quad 36/5\mathbf{R}$ $\quad 144/7\mathbf{R}$

3 Consider the three electron jumps described below for the hydrogen atom.

A. $n = 2$ to $n = 1$

B. $n = 3$ to $n = 2$

C. $n = 4$ to $n = 3$

(*a*) The photon emitted in which transition, *A*, *B*, or *C*, will have the longest wavelength?

(*b*) For which transition will the electron experience the largest change in orbit radius?

4 The Rydberg formula for one-electron systems other than H is

$$\frac{1}{\lambda} = RZ^2\left(\frac{1}{n_L^2} - \frac{1}{n_U^2}\right) \qquad \text{where } Z \text{ is the nuclear charge}$$

(*a*) A "Balmer" series ($n_L = 2$) for the He^+ ion ($Z = 2$) should be observed shifted toward the (infrared, ultraviolet) compared to that of the hydrogen atom.

(*b*) The wavelength of the lowest-energy Balmer line for the hydrogen atom is 656.3 nm. What is the wavelength of the lowest-energy Balmer line for He^+?

REFERENCES FOR FURTHER READING

1 Gamow, George, "One Two Three . . . Infinity," chap. VI, New American Library, New York, 1954.

2 ———, "The Atom and Its Nucleus," chap. 4, Prentice-Hall, Inc., Englewood Cliffs, N.J., 1961.

3 Garrett, A. B., The Flash of Genius, 9: The Bohr Atomic Model: Niels Bohr, *J. Chem. Educ.*, **39,** 534 (1962).

4 Davis, Jeff C., Jr., Introduction to Spectroscopy: Part V. Emission Spectra, *Chemistry*, **48,** 5, December (1975).

5 Kragh, H., Chemical Aspects of Bohr's 1913 Theory, *J. Chem. Educ.*, **54,** 208 (1977).

WAVE MECHANICS AND ATOMS

3

INTRODUCTION

3-1 More successful than the Bohr theory in explaining the structure and spectra of atoms is the mathematically sophisticated theory known as wave mechanics or quantum mechanics. Fortunately for us (the mathematically unsophisticated), one of the beauties of the theory is that it yields a treasure of simple rules and pictures from which we can gain an appreciation of the modern concepts of atomic structure.

THE SCHRÖDINGER EQUATION AND THE HYDROGEN ATOM

3-2 We begin by looking briefly at the equation formulated by Schrödinger in 1926, upon which almost everything we shall deduce about the behavior of the electron in the atom is based. For the simplest atomic system, the hydrogen atom, it is written

$$\frac{-h^2}{8\pi^2 m}\nabla^2\psi + V\psi = E\psi$$

This weighty equation (which we need not understand in detail at this point) is simply a symbolical way of stating that the total energy of the hydrogen atom, E, is the sum of the potential energy (the term containing V) and its kinetic energy (disguised in the first term). Some of the symbols we have already met: h is Planck's constant, and m is the mass of the electron.

Since this equation was derived by Schrödinger from the classical equations governing the behavior of waves, it is known as the Schrödinger wave equation, hence the name "wave mechanics."

Like most equations, this one contains "unknowns" which must be solved for, denoted by the Greek letter ψ (psi, pronounced "sigh"); but unlike most equations, there are *many* ψ's which are acceptable solutions.† Finding them is not an easy task, but accommodating mathematicians have done this for us. Once found, the ψ's, which depend on the coordinates of the electron and are thus *functions* (wave functions), can tell us almost all that we want to know about the behavior of the electron in the hydrogen atom.

Let us pause to consider the "almost." Because of what one might call the perversity of nature, the ψ's, though meaty with other information, cannot tell us the exact *position* of the electron in space at any specified time. Instead they tell us that the *probability* of finding the electron in some small chunk of space δv near the nucleus is related to $\psi^2\, \delta v$. The larger ψ^2 is in some section of space, the more likely the electron is to be found there. The probability interpretation is consistent with the idea that the electron is a particle, though described by a *wave* function.

Another interpretation perhaps more useful to chemists arises from the fact that in classical physics the intensity (photon density) of a light beam described by a wave function ϕ is related to ϕ^2. Hence we say that the magnitude of ψ^2 in some small element of space is a measure of the electron density there. According to this interpretation, the electron is smeared out in space, its density being greatest in those places where the corpuscular electron is likely to be found. In a sense the electron may now be regarded as a diffuse cloud rather than a small, discrete individual.

†The idea of an equation with many satisfactory solutions should not be new. For example, recall the simple trigonometric equation

$$\sin \alpha = 0$$

for which $\alpha = 0, \pi, 2\pi, 3\pi, \ldots$ or, in general, $\alpha = n\pi$, where $n = 0, \pm 1, \pm 2, \ldots$, are all solutions.

To distinguish these new pictures from the old, well-defined Bohr orbits, we name the ψ functions *orbitals* or *atomic orbitals*. The square of the orbital, ψ^2, defines the distribution of electron density in the orbital.†

The size and shape of the orbital cloud depend on which of the functions ψ we are considering. As a consequence of solving the Schrödinger equation, each of the orbital ψ's has associated with it three characteristic interrelated quantum numbers called n, l, and m_l. These arise naturally from the Schrödinger equation in much the same way as the integer n results from the general solution of the simple trigonometric equation in the footnote on page 22, and they are by no means assumptions, as were the quantum numbers of Bohr.

The principal quantum number n determines the size of the orbital cloud and also governs the allowed energy levels in the atom. n may assume the values 1, 2, 3, 4, . . . (any integer but not zero).

The azimuthal quantum number l determines the shape of the orbital cloud, and for any given value of n it may assume the values 0, 1, 2, 3, . . . , $n - 1$, that is, all integral values from zero up to a maximum of $n - 1$. For example, if $n = 4$, the l values associated with this n are 0, 1, 2, and 3. The different l values correspond to different angular momenta of the electron as it moves around the nucleus.

The quantum number m_l is related to the *orientation* of the orbital cloud in space. For a given value of l, m_l may take on values from $-l$ to $+l$ in integral steps. For example, for $l = 2$, m_l can be $-2, -1, 0, +1, +2$. This means that for any l there can be $2l + 1$ different values of m_l, or $2l + 1$ orbital clouds differing from one another in their orientation in space.

The interrelated quantum numbers serve in a sense as "addresses" for electrons; n parallels the floor number of a large apartment building, l the suite number of each floor, and m_l the room within the suite. One way to label these different locations is with the symbol ψ_{nlm}. For $n = 3$ (third floor) we

†Although the orbital is, strictly speaking, the ψ function, chemists frequently sketch the orbital cloud pictures associated with ψ^2 and refer to these as orbitals.

have:

ψ_{300}	$\psi_{311}, \psi_{310}, \psi_{31-1}$	$\psi_{322}, \psi_{321}, \psi_{320}, \psi_{32-1}, \psi_{32-2}$
Suite 0	Suite 1	Suite 2
(One room)	(Three rooms)	(Five rooms)
$l=0$	$l=1$	$l=2$
$m_l=0$	$m_l=1,0,-1$	$m_l=2,1,0,-1,-2$

Although the labels ψ_{nlm} are useful in some applications, chemists have proposed a less clumsy nomenclature for the orbitals, which, once mastered, is easily associated with the shapes of the orbital clouds.

All orbitals with $l=0$ are called *s* orbitals and all are associated with spherical cloud pictures (Fig. 3-1). The nucleus of the atom is at the origin of the coordinate system. Since wave mechanics indicates that there is a finite (though very small) probability of finding the electron even a couple of miles from the nucleus, the cloud pictures should have hazy rather than sharp boundaries. Instead of drawing diffuse clouds, we usually abbreviate them with boundary surfaces containing perhaps 95 percent of the electron density. On these boundary surfaces the electron density is constant.

FIGURE 3-1
Comparative boundary surfaces for 1*s* and 2*s* orbitals.

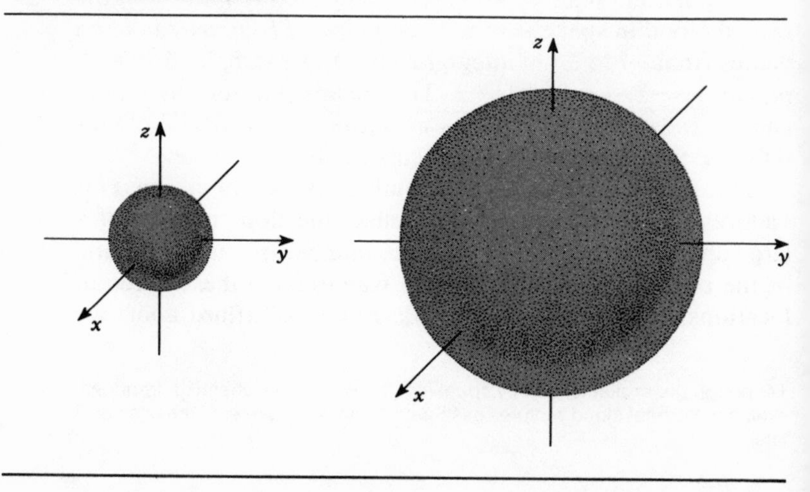

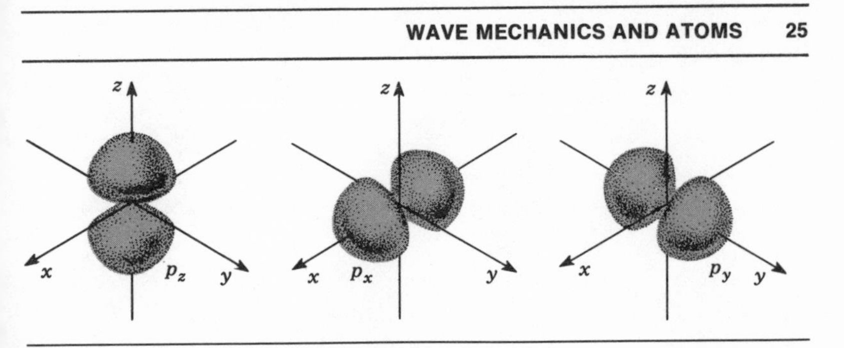

FIGURE 3-2
The *p* orbitals.

n appears in the orbital name as an integer in front of the l value. Thus ψ_{100} and ψ_{200} are represented in Fig. 3-1 as $1s$ and $2s$ orbitals. Note that as n gets larger, the boundary surface gets larger, paralleling the increase in radius with n predicted by Bohr theory.

When $l = 1$, the orbital is called a p orbital. Since for $l = 1$, $m_l = 1, 0, -1$, there are three kinds of p orbitals for a given n, differing from one another in orientation. All are associated with the same shape (Fig. 3-2); their boundary surfaces resemble dumbbells, and in contrast to the spherically symmetric s orbitals, the p orbitals possess directional properties. The three are named p_x, p_y, and p_z because their lobes of maximum probability lie along the x, y, and z axes, respectively. An electron in a p orbital may be found in either lobe with equal probability. Both lobes constitute one p orbital.

When $l = 2$, the orbital is called a d orbital. For a given n, there are five different m_l values and consequently five different d orbitals. Three of these are shown in cross section in Fig. 3-3. d_z^2 has a shape somewhat different from the other four; it is centered around the z axis, the largest part of the volume being somewhat p-orbital–shaped with a doughnut-shaped cloud encircling its middle. The other four d orbitals all have a cloverleaf shape with four lobes each. The $d_{x^2-y^2}$ orbital lies in the xy plane, with its four lobes coinciding with the x and y axes; d_{xy} also lies in the xy plane, but with its lobes pointed between the axes; d_{xz} and d_{yz} lie in the xz and yz

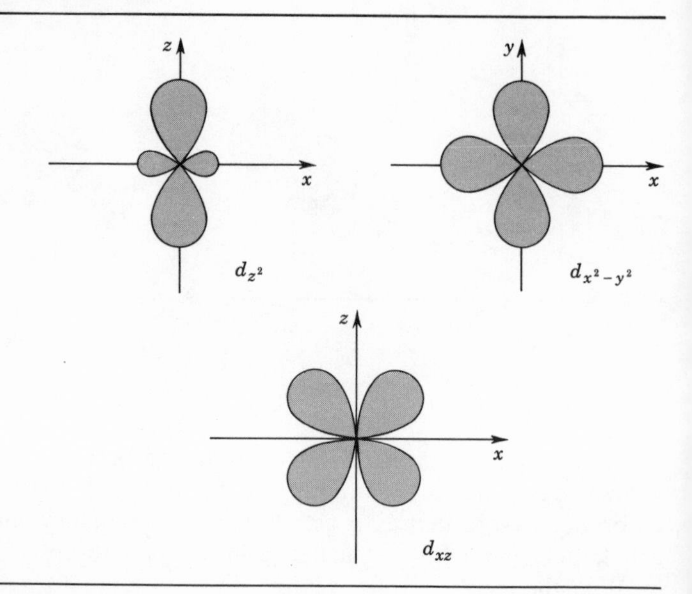

FIGURE 3-3
Three of the five *d* orbitals in cross section. These are shown in three dimensions in Fig. 6-5.

planes, respectively, and like d_{xy} have their lobes pointed between the axes.

For *l* values of 3, 4, and 5 the orbitals are named *f*, *g*, and *h*, respectively. The shapes of *f*, *g*, and *h* orbitals are less important in chemistry than those of *s*, *p*, and *d* orbitals. Hence we have not attempted to draw these.

The names of some orbitals that will be of interest to us are listed in Table 3-1.

In later arguments we will need to know how the electron density varies inside these boundary diagrams. Is the density constant everywhere inside? Wave mechanics says No! Figure 3-4 shows a plot of the calculated density ψ^2 at a distance *r* from the nucleus for the 1*s*, 2*s*, 2*p*, and 3*d* orbitals. For 1*s* and 2*s* orbitals, the electron density is highest near the nucleus and tapers off with distance. In contrast, both 2*p* and

TABLE 3-1
Some Quantum Numbers and Corresponding Orbitals

n	l	ORBITAL NAMES	$2l+1$ NUMBER OF m_l VALUES OR NUMBER OF KINDS OF EACH ORBITAL	FULL NAMES OF ORBITALS
1	0	1*s*	1	1*s*
2	0	2*s*	1	2*s*
	1	2*p*	3	$2p_x, 2p_y, 2p_z$
3	0	3*s*	1	3*s*
	1	3*p*	3	$3p_x, 3p_y, 3p_z$
	2	3*d*	5	$3d_{z_2}, 3d_{x^2-y^2}, 3d_{xy}, 3d_{xy}, 3d_{yz}$

3*d* orbitals have electron densities of zero at the nucleus, rising to a maximum and then decreasing with distance. Note that the electron density in a 3*d* orbital is very diffuse, spread out over a large distance.

The *p* orbitals, unlike *s* orbitals, have a plane of zero electron density, i.e., a so-called *nodal plane* separating the two lobes.† For p_z, for example, the *xy* plane is the nodal plane. Existence of these planes will be important in Chap. 4 when we try to classify different types of bonds.

With respect to energy, wave mechanics indicates that the electron in the hydrogen atom is governed by the same relationship that Bohr derived:

$$E_n = \frac{-2\pi^2 e^4 m}{n^2 h^2}$$

The lowest-energy state (the most stable state) of the electron in the hydrogen atom is that for which $n = 1$, corresponding to

†Often students are troubled by the seemingly paradoxical statement that an electron in a *p* orbital occupies each lobe with equal probability, although the lobes are separated by a nodal plane. The question arises: How does the electron get from one lobe to another? The answer we prefer is that the plane has mathematical, not physical, significance; i.e., it has zero thickness. It makes no sense to speak of finding a particle in that plane. In a slice of finite thickness on either side of or including that plane there is a nonzero probability of finding the electron. For more on this idea, see Ref. 12.

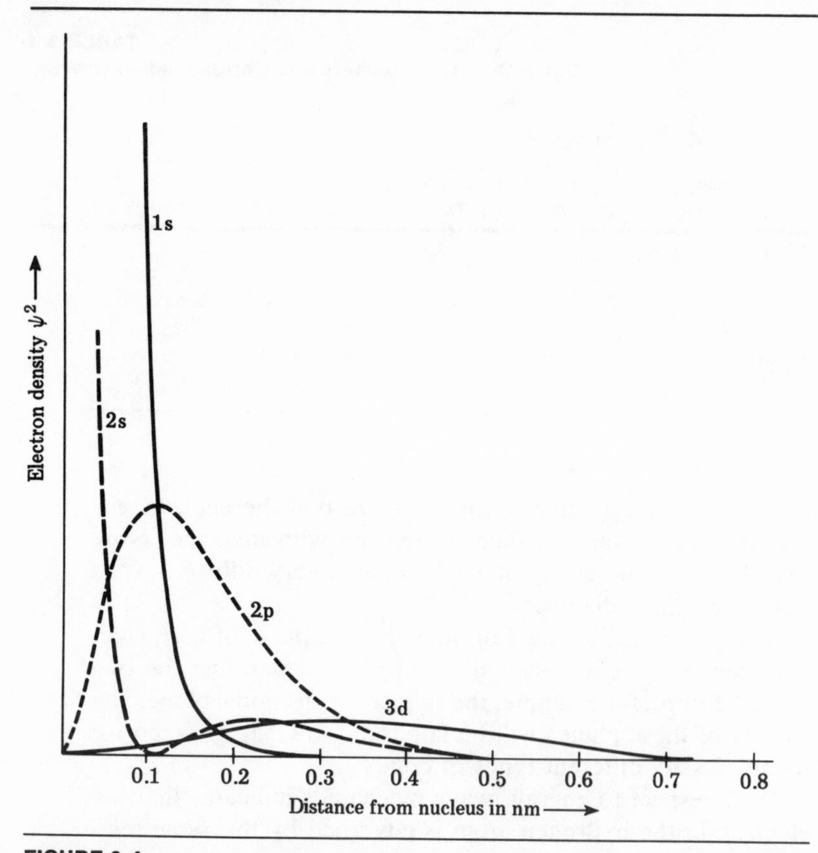

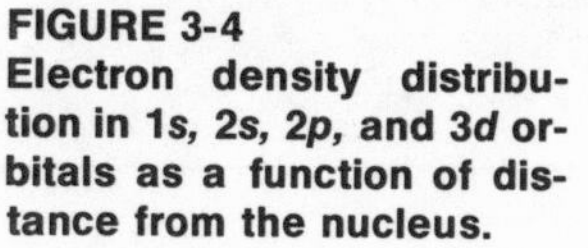

FIGURE 3-4
Electron density distribution in 1*s*, 2*s*, 2*p*, and 3*d* orbitals as a function of distance from the nucleus.

occupation of the 1*s* orbital. Higher-energy states (excited states) are those for which n is greater than 1. The spacings of these higher-energy states relative to the lowest (ground) state are shown on the ordinate of Fig. 3-6. For the hydrogen atom, the energy of the orbitals depends only upon the principal quantum number n (and not on l or m_l), and consequently the 2*s* and the three 2*p* orbitals all have the same energy. The

$n = 2$ level is said to have a fourfold orbital *degeneracy*; four different orbitals belong to this same energy level. Similarly the $n = 3$ level is ninefold degenerate, since the single 3*s* orbital, the three 3*p* orbitals, and the five 3*d* orbitals all have exactly the same energy.

Wave mechanics, like Bohr theory, says that the *emission spectrum* of the hydrogen atom results when electrons are excited to levels of higher *n* and subsequently fall back down to lower states, each downward jump being accompanied by the emission of a quantum of light, a photon. Here though, the electron is changing its *orbital* rather than its *orbit*.

MANY-ELECTRON ATOMS

3-3 The Schrödinger equation has been solved exactly only for one-electron systems, primarily because, even in a simple atom like helium ($Z = 2$), the repulsion between the two electrons makes the potential energy term V tremendously complicated. Thus the orbitals, quantum numbers, and pictures that we have discussed are not rigorously true even for He. However, there are many experimental facts which indicate that in larger atoms something quite like the quantum numbers n, l, and m_l governs the behavior of the electrons. This in turn implies that states like the *s*, *p*, *d*, *f*, . . . atomic orbitals exist and are occupied by these many electrons; consequently, we extrapolate the one-electron orbital results to bigger atoms. Justification for doing this is that it works!

Before we discuss the extrapolation (called *Aufbau*, or build-up), we state the existence of a fourth quantum number for the electron, namely, m_s, its "spin" quantum number. Unlike n, l, and m_l, m_s did not evolve from the Schrödinger equation but was proposed by experimentalists shortly before the Schrödinger equation was postulated. Uhlenbeck and Goudsmit found that a great deal of spectroscopic data could be explained *if it were postulated* that the electron is able to spin in one of two possible directions about an arbitrary axis through its center (Fig. 3-5). According to the direction of spin we assign to the electron $m_s = +\frac{1}{2}$ or $-\frac{1}{2}$. Usually we refer to these as α or β spin, and in comparing two different electrons

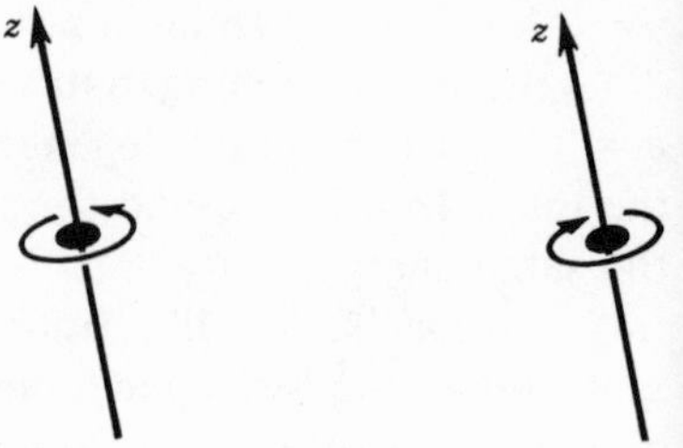

FIGURE 3-5
Electron spin around an arbitrary z axis.

we use the notation ↑↑ to mean same or unpaired spins (either $\alpha\alpha$ or $\beta\beta$) and ↑↓ to mean spins opposite or paired ($\alpha\beta$).

The existence of spin quantum numbers causes us to add to our collection of rules the *Pauli exclusion principle*, which states that each orbital can be occupied by at most two electrons, and then only if their spins are opposite or paired ($\alpha\beta$). For example, consider the hydride ion H^- (present in compounds of certain metals with hydrogen, such as LiH). Starting with the *neutral* hydrogen atom:

$$H\,(Z = 1)\!:\; 1s^1$$

(meaning that the charge Z on the nucleus is +1 and that there is one electron in the 1*s* orbital), we add one electron more to the 1*s* orbital to make

$$H^-(Z = 1)\!:\; 1s^2$$

or pictorially to show the electron pairing H^-: $\overset{1s}{(↑↓)}$

No more electrons can be accommodated in the 1*s* orbital according to the Pauli principle.

The helium atom is isoelectronic (having the same number of electrons) with the hydride ion but has two protons in the nucleus. Because of this increased nuclear charge we should expect the two electrons in He to be more tightly bound to the atom than those in H^-. It would require more energy to remove an electron from the vicinity of the nucleus, or in other words the $1s$ orbital of He would be more stable than that of H. The ground-state configuration of He would still be

$$\text{He}(Z = 2)\text{: } 1s^2 \qquad \text{or pictorially} \qquad \text{He: } \overset{1s}{\textcircled{\uparrow\downarrow}}$$

For the lithium atom, with three electrons, after pairing two electrons in the $1s$ orbital, we are faced with the problem of where the third electron will be. If the orbital energy levels in the Li atom were the same as those in H, it would not matter whether this last electron were assigned to the $2s$ or to a $2p$ orbital, since for H all of these possess the same energy.

Let us reexamine Fig. 3-4. Note that the electron density close to the nucleus is greater for a $2s$ than for a $2p$ orbital. Because of this we say that a $2s$ electron "penetrates" closer to the nucleus than a $2p$ electron. In the lithium atom an electron in a $2s$ orbital penetrates the electron cloud of the $1s$ electrons and sees more of the positive charge on the nucleus than an electron in a less penetrating $2p$ orbital, which sees a nuclear charge shielded by the two $1s$ electrons. Hence the $2s$ electron is more strongly attracted, is more difficult to remove from the atom (ionize), and is in a stabler state (lower in energy) than an electron in a $2p$ orbital.

Because of this penetration and shielding effect, part but not all of the orbital energy degeneracy evident in the hydrogen atom is removed. The three $2p$ orbitals all have the same penetrating power and consequently the same energy in a many-electron atom.

In general, for any given principal quantum number n, s orbitals are lower in energy than p orbitals, which are lower than d orbitals, etc., since the penetrating powers vary as follows:

$$\text{Most penetrating} = s > p > d > f > g > h \cdots$$

Figure 3-6, constructed from theoretical and experimental considerations, shows how the orbital energies may change in neutral atoms as the size of the atom increases. Note that, as the atom gets larger, in certain cases the principal quantum number loses control of the energy, and the orbital-energy lines cross; for example, $4s$ becomes lower than $3d$ because of the high penetrating power of the s orbital even though, according to the principal quantum number, $3d$ should be lower in energy. We shall discuss this crossover in detail in Sec. 3-4.

For many of the lighter elements a rough qualitative rule of orbital energies for atoms in their ground states is:

$$\text{Lowest} = 1s < 2s < 2p < 3s < 3p < 4s < 3d < 4p < 5s < 4d \cdots$$
$$n + l = 1 \quad 2 \quad 3 \quad 3 \quad 4 \quad 4 \quad 5 \quad 5 \quad 5 \quad 6$$

The $n + l$ values are listed as a mental crutch. Usually the energy varies first as $n + l$, and within a group of identical $n + l$ the orbital with lowest n is lowest in energy.

FIGURE 3-6
Relative orbital energies of neutral atoms. Energies of the H atom appear on the ordinate.

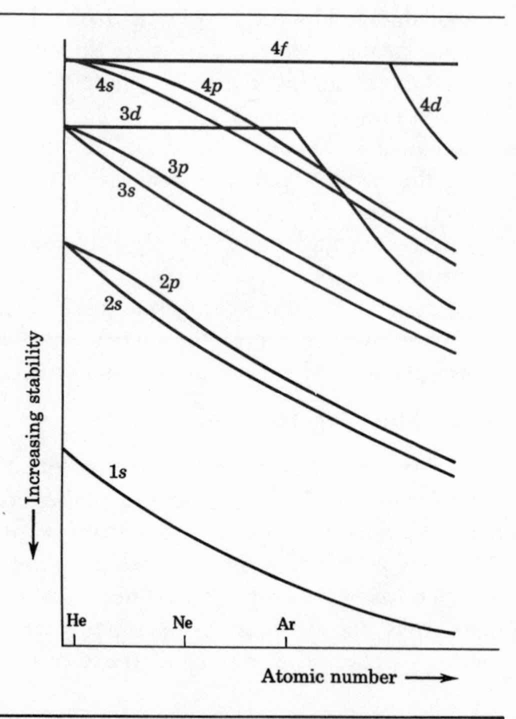

If we return now to the *Aufbau* principle and the question of Li, its ground-state configuration is obviously

		1s	2s
Li ($Z = 3$):	$1s^2\,2s^1$	⇅	↑

and those of others across the periodic table:†

		1s	2s	2p		
Be ($Z = 4$):	$1s^2\,2s^2$	⇅	⇅	○	○	○
B ($Z = 5$):	$1s^2\,2s^2\,2p^1$	⇅	⇅	↑	○	○
C ($Z = 6$):	$1s^2\,2s^2\,2p^2$	⇅	⇅	↑	↑	○

For carbon we introduce *Hund's rule*, which in essence says that electrons prefer to occupy separate orbitals with their spins the same provided that the energies of the separate orbitals are the same or similar. Since all three of the $2p$ orbitals have the same energy, the two $2p$ electrons in carbon are quite happy to occupy separate homes, say arbitrarily $2p_x$ and $2p_y$. To pair in the same orbital requires a considerable expenditure of energy by the electron, since like charges repel one another and the first electron occupying an orbital repels the second.

Why then do not the two electrons in helium enter one into the $1s$ orbital and the other into the $2s$ orbital? In a sense, the second electron must decide which is greater: the energy lost by occupying the less stable $2s$ orbital or the energy lost in overcoming the repulsion of the $1s$ electron. Since the energy separation between the $1s$ and $2s$ levels is relatively large (see Fig. 3-6), the lesser of these two evils is the latter.

Continuing across the periodic table (App. A), we have

		1s	2s	2p		
N ($Z = 7$):	$1s^2\,2s^2\,2p^3$	⇅	⇅	↑	↑	↑
O ($Z = 8$):	$1s^2\,2s^2\,2p^4$	⇅	⇅	⇅	↑	↑
F ($Z = 9$):	$1s^2\,2s^2\,2p^5$	⇅	⇅	⇅	⇅	↑
Ne ($Z = 10$):	$1s^2\,2s^2\,2p^6$	⇅	⇅	⇅	⇅	⇅

†A periodic table is given in App. A.

Between the $2p$ and the $3s$ levels there is a considerable energy gap (Fig. 3-6). In fact, for low Z there are at least four groups of levels characterized by being close to one another in energy and separated by a wider gap from other groups. These are $1s$ (a group in itself); $2s$, $2p$; $3s$, $3p$; and $4s$, $3d$, $4p$. Each group makes up what may be called an energy shell, similar to Bohr's shells around the atom. Filling one of these shells with its quota of electrons corresponds to attaining the peculiar stability associated with the noble gases. The first four noble gases, their energy shells, and the capacities of these (in parentheses) are:

	(2)	(8)	(8)	(18)
He ($Z = 2$):	$1s^2$			
Ne ($Z = 10$):	$1s^2$	$2s^2\,2p^6$		
Ar ($Z = 18$):	$1s^2$	$2s^2\,2p^6$	$3s^2\,3p^6$	
Kr ($Z = 36$):	$1s^2$	$2s^2\,2p^6$	$3s^2\,3p^6$	$3d^{10}\,4s^2\,4p^6$

Orbital theory thus makes it reasonable that the period from sodium to argon should comprise only 8 members and that the period from potassium to krypton should include 18 elements.

By means of the orbital theory the entire periodic table (App. A) is given a sound theoretical basis. For example, all the members of one family (vertical column) have similar external electronic configurations; these are responsible for the similarity in chemical behavior of most members of the family. In the two following illustrations the similar external parts are in bold-faced type.

THE ALKALI METALS (GROUP I)

Li ($Z = 3$):	(He core)^2 $\mathbf{2s^1}$
Na ($Z = 11$):	(Ne core)^{10} $\mathbf{3s^1}$
K ($Z = 19$):	(Ar core)^{18} $\mathbf{4s^1}$
Rb ($Z = 37$):	(Kr core)^{36} $\mathbf{5s^1}$

All the alkali metals readily lose one electron (the outermost s electron) to form singly positive ions Li^+, Na^+, K^+, Rb^+ with noble-gas configurations, but it is extremely difficult to ionize them further to doubly positive species.

THE HALOGENS (GROUP VII)

F ($Z = 9$): (He core)^2 $\mathbf{2s^2\ 2p^5}$
Cl ($Z = 17$): (Ne core)^{10} $\mathbf{3s^2\ 3p^5}$
Br ($Z = 35$): (Ar core)^{18} $3d^{10}$ $\mathbf{4s^2\ 4p^5}$

The halogens readily *gain* one electron (completing their outermost *p* levels) to form singly negative ions F^-, Cl^-, Br^-. These have the structure of the adjacent noble gas, Ne, Ar, and Kr, respectively. Once the noble-gas configuration is attained, there is little or no tendency to gain more electrons.

The lengths of the fourth and fifth periods (10 extra members as compared to the second and third periods) are caused by the large number of electrons (10) necessary to fill up the $3d$ and $4d$ levels, respectively.

The 24 "extra" members of the sixth period result from the filling of the $4f$ and $5d$ levels (14 and 10 electrons, respectively).

ATOMS AND IONS OF THE FIRST TRANSITION SERIES

3-4 Application of the *Aufbau* principle to the ground states of atoms from helium ($Z = 2$) to argon ($Z = 18$) is rather straightforward; there are no exceptions to the general rules outlined in the previous section. After potassium [(Ar core)^{18} $4s^1$] and calcium [(Ar core)^{18} $4s^2$], the $3d$ level gradually fills up as we progress across the first transition series from scandium [(Ar core)^{18} $4s^2$ $3d^1$] to zinc [(Ar core)^{18} $4s^2$ $3d^{10}$] with but two apparent exceptions at chromium [(Ar core)^{18} $4s^1$ $3d^5$] and copper [(Ar core)^{18} $4s^1$ $3d^{10}$], both of which have an incomplete $4s$ level. The latter two cases may be understood if we examine in Fig. 3-7 a blowup of a portion of Fig. 3-6.

In region *A*, covering elements from H to Ar, the ground-state configuration of the atom includes no $3d$ or $4s$ electrons; thus the lines in Fig. 3-7 represent energies of electrons ex-

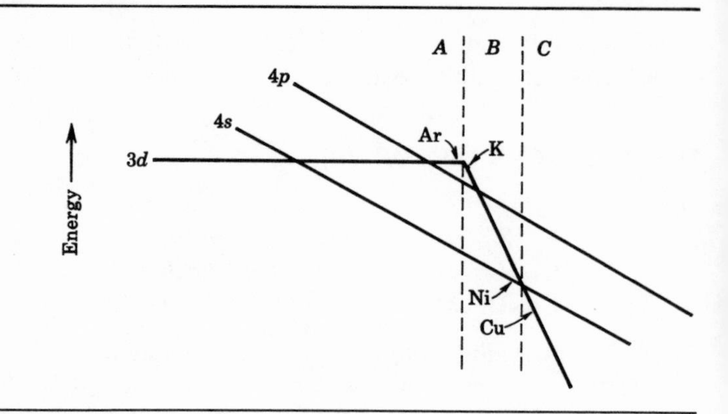

FIGURE 3-7
Orbital energies for atoms near the first transition series.

cited from the core to the 4*s* or 3*d* levels. Now consider what an excited electron sees at the core as we go from H to Ar. Simultaneously we increase the nuclear charge and add electrons to the core. A 4*s* electron, being very penetrating, sees this increased charge and becomes more tightly bound, and the 4*s* level drops in energy. A 3*d* electron, only weakly penetrating, sees little change in the nuclear charge as the core grows larger; its energy or stability remains relatively constant.

After argon we arrive in region *B*, where 4*s* electrons are added externally as the nuclear charge increases. The 4*s* electrons have a high density in the same region of space as the 3*d* electron. Thus, being no longer between the 3*d* electron and the nucleus, the added 4*s* electrons do not shield the 3*d* electron from the increased nuclear charge, and the 3*d* level now begins to drop in energy because of increased nuclear attraction. The drop becomes so fast as electrons are added to the *d* level that 3*d* eventually becomes lower in energy than 4*s*, the crossover point defining region *C*. Experimentally it seems that the crossover occurs just after the nickel atom [(Ar core)^{18} $4s^2$ $3d^8$], so that for the next atom, Cu, 3*d* is considerably lower in energy than 4*s*, and the 11 electrons outside the Ar core take the configuration $3d^{10}$ $4s^1$.

The apparent exception Cr ($4s^1\ 3d^5$) still remains, for, in region *B*, 4*s* is lower than 3*d*. For Cr, however, the 3*d* and 4*s* levels are close enough that Hund's rule prevails, and the six electrons apparently prefer to occupy singly the 4*s* and five 3*d* orbitals with spins unpaired rather than pairing in a $4s^2\ 3d^4$ configuration. The ground-state configurations of all atoms from argon to zinc are shown in Table 3-2.

For the ground states of *ions* of these elements no discontinuities occur in the filling of the *d* level. Normally the atoms lose at least two electrons in the ionization process, and the configurations of the ions reflect a gradual filling of the *d* orbitals (Table 3-2). It seems puzzling that for all these ions the 4*s* rather than 3*d* electrons are gone, particularly since for most of the neutral atoms the 4*s* electrons are lower in energy, more stable than the 3*d*. The experimental facts, however, are unequivocal and indicate that in the ions the 3*d* level must drop much earlier than it does in the neutral atoms. It seems quite probable that even in neutral atoms shielding of nuclear charge by core electrons is not a one-to-one process and that in the metal ions complete loss of two shielding electrons is sufficient to allow the *d* electron to see the increasing nuclear charge and drop in energy sooner. Effects like this and the so-called anomalous behavior of neutral chromium and copper are

TABLE 3-2
Ground-state Configurations of Transition-metal Atoms and Ions

Ar	($Z = 18$):	$1s^2\ 2s^2\ 2p^6\ 3s^2\ 3p^6$		
K	($Z = 19$):	(Ar core)^{18} $4s^1$	K^+:	(Ar core)^{18}
Ca	($Z = 20$):	(Ar core)^{18} $4s^2$	Ca^{2+}:	(Ar core)^{18}
Sc	($Z = 21$):	(Ar core)^{18} $4s^2\ 3d^1$	Sc^{2+}:	(Ar core)^{18} $3d^1$
Ti	($Z = 22$):	(Ar core)^{18} $4s^2\ 3d^2$	Ti^{2+}:	(Ar core)^{18} $3d^2$
V	($Z = 23$):	(Ar core)^{18} $4s^2\ 3d^3$	V^{2+}:	(Ar core)^{18} $3d^3$
Cr	($Z = 24$):	(Ar core)^{18} $4s^1\ 3d^5$	Cr^{2+}:	(Ar core)^{18} $3d^4$
Mn	($Z = 25$):	(Ar core)^{18} $4s^2\ 3d^5$	Mn^{2+}:	(Ar core)^{18} $3d^5$
Fe	($Z = 26$):	(Ar core)^{18} $4s^2\ 3d^6$	Fe^{2+}:	(Ar core)^{18} $3d^6$
Co	($Z = 27$):	(Ar core)^{18} $4s^2\ 3d^7$	Co^{2+}:	(Ar core)^{18} $3d^7$
Ni	($Z = 28$):	(Ar core)^{18} $4s^2\ 3d^8$	Ni^{2+}:	(Ar core)^{18} $3d^8$
Cu	($Z = 29$):	(Ar core)^{18} $3d^{10}\ 4s^1$	Cu^{2+}:	(Ar core)^{18} $3d^9$
Zn	($Z = 30$):	(Ar core)^{18} $3d^{10}\ 4s^2$	Zn^{2+}:	(Ar core)^{18} $3d^{10}$

results of highly complex interelectronic forces and cannot as yet be predicted accurately. We can only (as in much of chemistry) search the experimental facts and try to propose reasons for the observed behavior.

EXERCISES

1 Which of the following orbitals do and which don't "make sense" according to wave mechanics? Explain your answers.

$2d$ $6h$ $7g$ $3f$

2 What is the maximum number of electrons that can be accommodated in

all the $6g$ orbitals?
all the $7s$ orbitals?
all the $8f$ orbitals?
all the orbitals with $n = 5$?

3 Write the complete ground-state electronic configurations of the first three members of the nitrogen family: N ($Z = 7$), P ($Z = 15$), and As ($Z = 33$), showing all unpaired electrons. Underline the parts responsible for their similarity in chemical behavior.

4 Without referring to the text write ground-state electronic configurations for the following atoms, applying Hund's rule whenever appropriate:

Mg ($Z = 12$) Mo ($Z = 42$)
Si ($Z = 14$) Xe ($Z = 54$)
Ca ($Z = 20$) I ($Z = 53$)
Ni ($Z = 28$) Cs ($Z = 55$)

5 List all ions of the first transition series with d^5 or d^{10} ground-state configurations.

6 In general, the first ionization energies (energies required to remove one electron from the gaseous neutral atom) increase as we go across the second period (see below). Why? However, boron and oxygen exhibit anomalous behavior; their ionization energies do not fit into this smooth trend. Examine the electronic configuration of these elements and suggest a qualitative reason why.

	Li	Be	B	C	N	O	F	Ne
I(kJ/mol)	519	895	799	1092	1397	1314	1699	2084

7 The energy required to remove the outermost electron from an Sr atom is 548 kJ/mol; removal of a second electron requires almost twice as much energy, 1058 kJ/mol. In comparison, the first ionization energy of Rb is 402 kJ/mol; removal of the second electron requires almost seven times as much energy, 2640 kJ/mol. Suggest a qualitative reason for these observations on the basis of the electronic configurations of Sr and Rb.

8 Beginning with 4*d*, predict the next five orbitals in increasing energy for a many-electron atom; i.e., continue the series illustrating the $n + l$ rule in Sec. 3-3. What would be the order of these levels on the hydrogen atom?

9 Theoretical calculations have yielded the following values for the radii of small isolated neutral atoms:

ATOMIC RADII, IN NANOMETERS

H							He
0.053							0.030
Li	Be	B	C	N	O	F	Ne
0.150	0.119	0.088	0.066	0.056	0.048	0.041	0.037

(*a*) Why does the helium atom, with 2 electrons, have a smaller radius than that of hydrogen?

(*b*) Why is the radius of the lithium atom so much larger than that of the hydrogen atom?

(*c*) Based on trends observed above, rearrange the atoms in these groups in order of decreasing radius:

(1) Al, Na, Ar, P

(2) Si, Sr, Ar, Ge

(3) Se, Cl, Sb, Ne
(4) Se, O, Te, S

REFERENCES FOR FURTHER READING

GENERAL

1 Heslop, R. B., and P. L. Robinson, "Inorganic Chemistry," 3d ed., chap. 4, Elsevier Publishing Company, Amsterdam, 1967.

2 Coulson, C., "Valence," 2d ed., chap. II, Oxford University Press, New York, 1961.

3 Maybury, R. H., The Language of Quantum Mechanics, *J. Chem. Educ.*, **39,** 367 (1962).

4 Slezak, J. A., The Concept of Quantum Numbers—A Simplified Approach, *J. Chem. Educ.*, **48,** 485 (1971).

ORBITAL SHAPES, ELECTRON DENSITIES, NODES

5 Cohen, I., The Shape of 2*p* and Related Orbitals, *J. Chem. Educ.*, **38,** 20 (1961).

6 Bordass, W. T., and J. W. Linnett, A New Way of Presenting Atomic Orbitals, *J. Chem. Educ.*, **47,** 672 (1970).

7 Ogryzlo, E. A., and G. B. Porter, Contour Surfaces for Atomic and Molecular Orbitals, *J. Chem. Educ.*, **40,** 256 (1963).

8 Berry, R. Stephen, Atomic Orbitals, *J. Chem. Educ.*, **43,** 283 (1966).

9 Powell, Richard E., Relativistic Quantum Chemistry: The Electrons and the Nodes, *J. Chem. Educ.*, **45,** 558, 790E (1968).

10 Cohen, I., and T. Bustyard, Atomic Orbitals—Limitations and Variations, *J. Chem. Educ.*, **43,** 187 (1966).

11 Cromer, Don T., Stereo Plots of Hydrogen-like Electron Densities, *J. Chem. Educ.*, **45,** 626 (1968).

12 Ellison, F. O., and C. A. Hollingsworth, The Probability Equals Zero Problem in Quantum Mechanics, or How Does an Electron Get from First to Second to Third without Touching Second? *J. Chem. Educ.*, **53,** 767 (1976).

13 Perlmutter-Hayman, Berta, Graphical Representation of Hydrogen-like Wave Functions, *J. Chem. Educ.*, **46,** 428 (1969).

14 Szabo, Attila, Contour Diagrams for Relativistic Orbitals, *J. Chem. Educ.*, **46,** 678 (1969).

15 Lambert, Frank L., Atomic Orbitals from Wave Patterns, *Chemistry*, **41,** 10, February (1968); **41,** 8, March (1968).

16 Mak, T. C. W., and W.-K. Li, Relative Sizes of Hydrogenic Orbitals and the Probability Criterion, *J.Chem. Educ.*, **52,** 90 (1975).

17 Friedman, H. G., Jr., G. R. Choppin, and D. G. Feuerbacher, The Shapes of *f* Orbitals, *J. Chem. Educ.*, **41,** 354 (1964).

18 Becker, Clifford, Geometry of the *f* Orbitals, *J. Chem. Educ.*, **41,** 358 (1964).

19 Ogryzlo, E. A., On the Shapes of *f* Orbitals, *J. Chem. Educ.*, **42,** 150 (1964).

ELECTRON CONFIGURATIONS AND TRENDS

20 Keller, R. N., Energy Level Diagrams and Extranuclear Building of the Elements, *J. Chem. Educ.*, **39,** 289 (1962).

21 Liebman, Joel F., Regularities and Relations among Ionization Potentials of Nontransition Elements, *J. Chem. Educ.*, **50,** 831 (1973).

22 Haight, G. P., Jr.,The Use of Tables of Data in Teaching—The Students Discover Laws about Ionization Potentials, *J. Chem. Educ.*, **44,** 468 (1967).

23 Sanderson, W. T., Ionization Energy and Atomic Structure, *Chemistry*, **46,** 12, May (1973).

24 Hochstrasser, Robin M., The Energies of the Electronic Configurations of Transition Metals, *J. Chem. Educ.*, **42,** 154 (1965).

PERIODIC TABLES

25 Hyde, J. F., A Newly Arranged Periodic Chart, *Chemistry*, **49,** 15, September (1976).

MOLECULES AND THE COVALENT CHEMICAL BOND

4

INTRODUCTION: MOLECULAR ORBITAL FORMATION

4-1 To define a molecule, we may say that it is a discrete group of atoms held together by chemical bonds. This in turn requires definition of a chemical bond, and this we shall try to supply in this chapter.

Just as the hydrogen atom is the simplest atom, the hydrogen molecule, H_2, is the simplest kind of molecule. What happens when a bond between two hydrogen atoms forms? To answer, let us imagine that two isolated H atoms, each with its electron in its ground-state 1*s* orbital, approach one another (Fig. 4-1).

Atoms, like humans, are continually seeking a state of greater security or stability, and we may imagine that these two atoms in their approach are looking one another over, counting the pros and cons of merger. As they get closer and closer, the 1*s* clouds containing the electrons begin to overlap. Each electron feels attracted to the approaching nucleus, and overlap increases. The two atomic orbitals merge into one bigger cloud called a molecular orbital (*MO*), and in it the electrons find that they are strongly attracted to both nuclei. When the repulsive forces between the positively charged nuclei have determined the position of closest approach, the merger halts. At this point the system of two nuclei and two electrons has attained a stability surprisingly greater than that of the two isolated atoms, and the molecule is born.

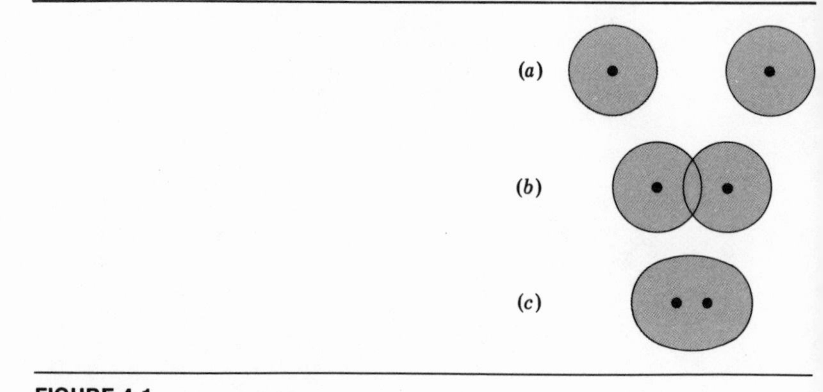

FIGURE 4-1
Molecular orbital formation from *s* orbitals. (*a*) Two isolated atomic orbitals; (*b*) overlap; (c) the molecular orbital boundary diagram.

Figure 4-2 shows graphically the energy changes just described. At large values of the internuclear distance R, the energy of the system is just that of two isolated hydrogen atoms. Arbitrarily we call this energy zero, so that any more stable state of the system will be described by a negative energy. As R decreases, the stability increases, largely because of the dual nuclear attractive forces acting on the electrons. At some point R_e (the equilibrium internuclear distance) the stability maximizes (as the energy minimizes), since, past this distance, at smaller R, the strong repulsion between the nuclei causes the curve to rise steeply.

An energy curve like this is sometimes called a potential well, because the behavior of the system it describes parallels the behavior of a rubber ball in a curved "well" of the same shape. If such a ball were placed at point A in Fig. 4-2, it would roll down to the bottom, oscillate about the point designated as R_e, and because of friction would eventually stop at R_e. Similarly, if enough energy is supplied to the molecule to stretch it (increasing its R and its energy to position A), the point describing the state of the molecule, the analog of the ball, will "roll back down into the well" to the equilibrium position when the stretching forces are removed. Similar arguments hold for compression of atoms in a molecule to an R value smaller than R_e (situation B in Fig. 4-2).

An energy D would be required to lift the rubber ball out of the well. Similarly, an energy D is sufficient to lift the molecule at its equilibrium distance out of the well. Above the well at C the energy of the system is just that of two isolated atoms; there is no bonding energy and the atoms may fly apart. Thus the energy D is called the *equilibrium dissociation energy* of the molecule, or the energy needed to break up the molecule into its component atoms in their ground states.

The peculiar stability of the electrons in the molecular orbital in H_2 is the fundamental reason for bond formation. In the molecular orbital (MO) the total electron density between the two nuclei is larger than the sum of the densities of the two overlapping atomic orbitals and acts as a negatively charged "glue" holding the positively charged nuclei together. In the

FIGURE 4-2
A potential-energy diagram for a diatomic molecule.

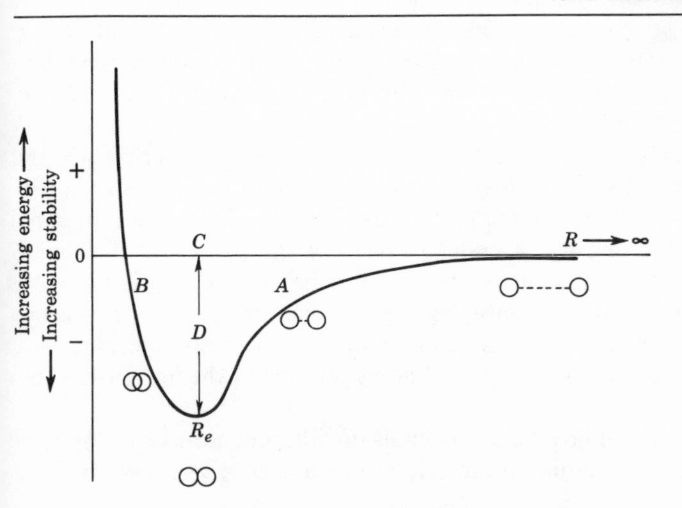

H_2 molecule the two electrons are shared equally by the two nuclei. This sharing corresponds to what chemists call a *covalent bond*.

This particular kind of molecular orbital is called a σ (Greek sigma) *MO*, and more explicitly may be designated σ_{1s}, since it was formed by merger of two 1*s* orbitals. The subscripts are necessary since, as we shall see in the next section, sigma-type *MO*s may be formed by appropriate merger of any type of orbital (2*s*, 2*p*, 3*d*, 4*f*, . . .). For hydrogen we may write the ground-state configuration of the *molecule* as

$$H_2: \quad \sigma_{1s}^2$$

Even in *MO*s the electrons are haunted by the Pauli exclusion principle, and consequently the two electrons in the ground-state *MO* of H_2 must have their spins paired.

Just as in the atom, there are available in the molecule many higher-energy, less-stable *MO*s, which may be constructed by merging 2*s* orbitals, 2*p* orbitals, etc. These will be discussed in detail in Sec. 4-11.

GENERAL PROPERTIES OF MOLECULAR ORBITALS

4-2 There are only three kinds of *MO*s that we need worry about in chemistry: the σ (sigma) *MO*s, usually associated with strong bonds, and their progressively weaker relatives, π (pi) *MO*s and δ (delta) *MO*s. To illustrate the differences between them, we shall consider construction of σ and π *MO*s from 2*p* orbitals.

In Chap. 3 we learned that *p* orbitals may be represented by dumbbell-shaped clouds lying along one of the *x*, *y*, or *z* axes, each *p* orbital having two lobes separated by a plane of zero electron density, a nodal plane. For the p_z orbital the nodal plane is simply the *xy* plane. It is important at this point to remember that the labeling of particular axes, *x*, *y*, or *z*, is arbitrary; we may label them and relabel them to suit our purpose. The electron in an orbital knows nothing of the letters *x*, *y*, or *z*.

Consider now two p_z orbitals on different atoms overlapping head on by pointing directly at one another, as shown in Fig.

4-3. The boundary diagram of the *MO* formed by the merger indicates that the electron density has piled up between the nuclei, and as a consequence the density in the outer lobes has decreased. Note that the *MO* thus formed has two nodal planes perpendicular to the internuclear axis, but *no nodal plane containing the internuclear axis*. This latter characteristic is the criterion for distinguishing a σ *MO* from all other *MO* types. Note that the σ *MO* formed from 1*s* orbitals had no such nodal plane.

When the *MO* is occupied by two electrons, we say that the atoms are joined by a σ bond. Sigma bonds are the strongest kind of covalent bond, with the largest bond-dissociation energies.

In contrast, two p_z orbitals on different atoms could approach one another sideways, as shown in Fig. 4-4. The boundary diagram after merger again shows increased electron density between the nuclei, and moreover the molecular orbi-

FIGURE 4-3
Sigma molecular orbital formation from *p* orbitals. (*a*) Two isolated p_z atomic orbitals; (*b*) head-on overlap; (*c*) the σ*MO* boundary diagram.

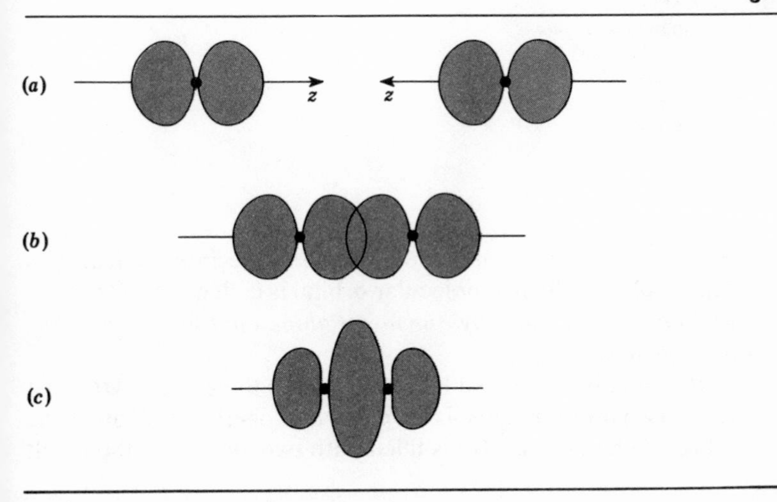

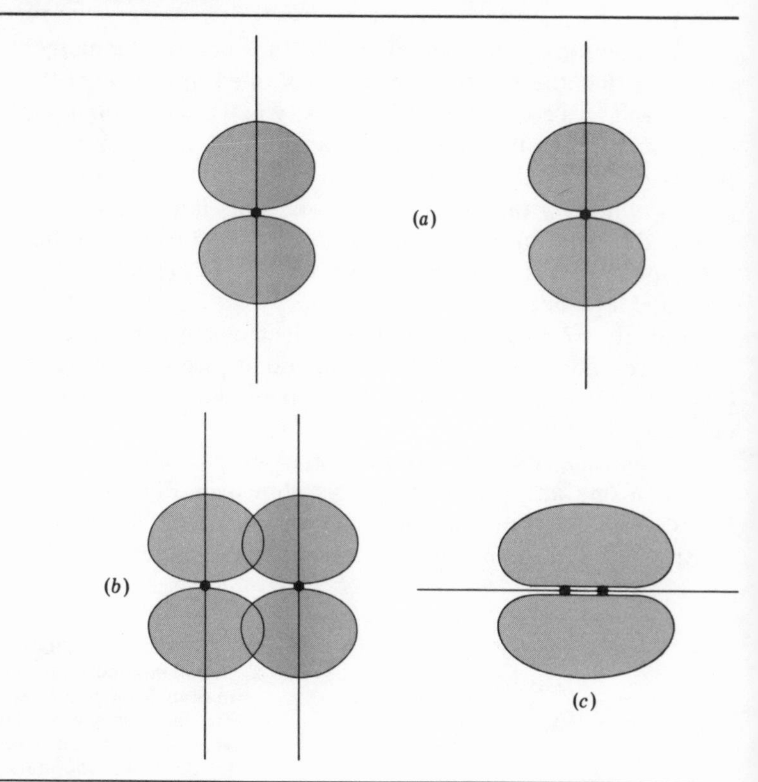

FIGURE 4-4
Pi molecular orbital formation from *p* orbitals. (*a*) Isolated p_z orbitals; (*b*) sideways overlap; (c) the π*MO* boundary diagram.

tal formed has a nodal plane *containing* the internuclear axis (the *xy* plane). Such a molecular orbital is called a π *MO*. All π *MO*s have *one and only one nodal plane containing the internuclear axis*.

Both the bean-shaped lobes are part of the same π *MO*, just as the two lobes of an isolated p_z orbital together constitute the orbital. When the π *MO* is filled with two electrons, the result is a π bond between the two atoms. Pi bonds are generally weaker than σ bonds.

Figure 4-5 shows the formation of a δ *MO* from two $3d$ orbitals overlapping face to face. The four sausagelike clouds all belong to the same *MO*. Superficially the δ *MO* resembles two perpendicular π *MO*s; however, two π *MO*s could accommodate four electrons, while the δ *MO*, even with four lobes, can accommodate only two electrons. Note the existence of *two nodal planes containing the internuclear axis* (the xz and the yz planes).

From the discussion thus far we can conclude that σ, π, and δ *MO*s have 0, 1, and 2 nodal planes containing the bond axis, respectively. The reader should satisfy himself that (1) *only* σ *MO*s can be constructed from s orbitals, (2) *only* σ and π *MO*s can be constructed from p orbitals, and (3) σ, π, and δ *MO*s can be constructed from d orbitals.

APPLICATION TO THE NITROGEN MOLECULE

4-3 A molecule considerably larger than H_2, yet similar in that it is homonuclear (same nuclei), is the nitrogen molecule N_2. As the first step in the analysis of its structure, we consider the ground-state electronic configuration of the nitrogen atom, ap-

FIGURE 4-5 Delta molecular orbital formation. (*a*) Isolated $d_{x^2-y^2}$ orbitals; (*b*) the δ *Mo*.

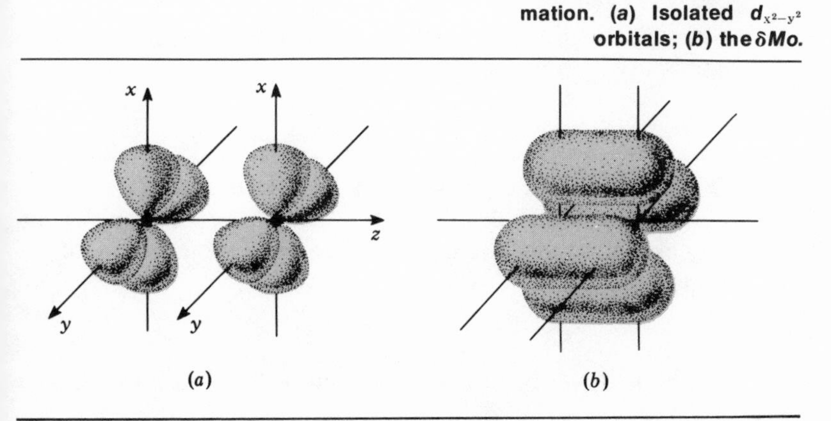

plying Hund's rule:

$$\text{N } (Z = 7)\text{: } 1s^2\ 2s^2\ 2p_x^1\ 2p_y^1\ 2p_z^1$$

In the simplest approach we can assume that, since the $1s$ and $2s$ orbitals on the nitrogen atoms are already filled with paired electrons, they will not contribute much to bonding and will not enter into *MO* formation. In this approximation we are treating them as "nonbonding" or "core" electrons, assuming that they will remain largely near their respective nuclei, forming "cores" like [N ($Z = 7$): $1s^2\ 2s^2$], around which the *MO*s will form by merger of the $2p$ orbitals. If we orient the coordinate systems of two nitrogen atoms a and b so that the two $2p_y$ orbitals are pointed at one another (Fig. 4-6*a*), then a σ *MO* will result from their overlap, accommodating the two $2p_y$ electrons (Fig. 4-6*b*). Upon sideways overlap of the $2p_x$ orbitals and the $2p_z$ orbitals, two π *MO*s (four bean-shaped clouds) will form surrounding the σ *MO*, as shown in Fig. 4-6*c*. Each of these π *MO*s will hold two electrons, so that the six electrons originally in $2p$ orbitals in the isolated atoms are now located two in a σ *MO* and two each in the π *MO*s. This combination of one σ bond and two π bonds corresponds to a triple bond between the atoms, so that in the classical line structure of N_2, N≡N, one line represents a σ bond and the other two π bonds.

The electronic configuration of the N_2 molecule may be written

$$N_2\text{:} \qquad 1s_a^2\ 1s_b^2\ 2s_a^2\ 2s_b^2\ \sigma_y^2\ \pi_x^2\ \pi_z^2$$

where we have abbreviated σ_{2p_y} as simply σ_y, since there is no ambiguity in doing so.

SOME SIMPLE HETERONUCLEAR MOLECULES

4-4 Diatomic molecules whose nuclei are not alike, i.e., heteronuclear molecules, can be treated by similar pairing schemes. Consider the molecule HF. The ground state of the F atom is

$$\text{F } (Z = 9)\text{:} \qquad 1s^2\ 2s^2\ 2p_x^2\ 2p_y^2\ 2p_z^1$$

The half-empty $2p_z$, when aimed at the $1s$ of a hydrogen

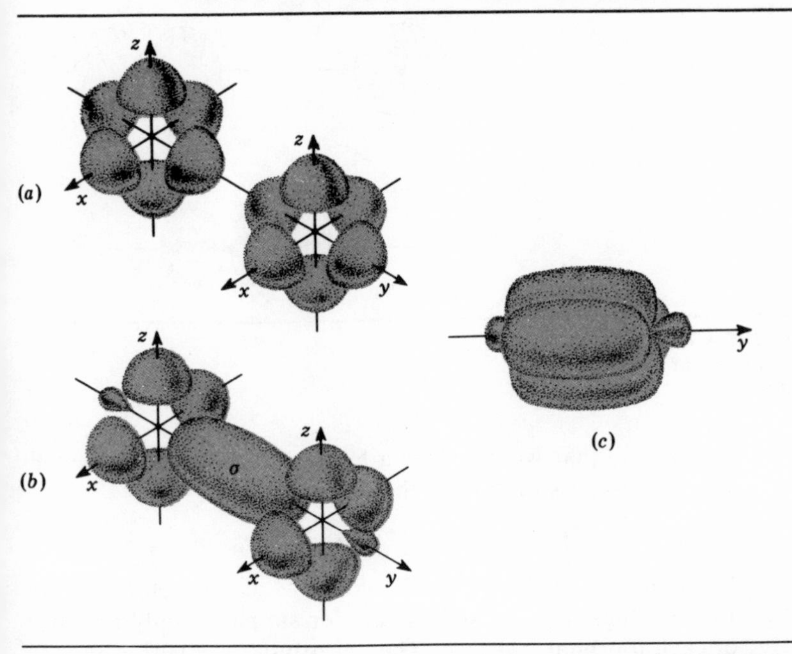

FIGURE 4-6
Molecular orbitals in the nitrogen molecule. (*a*) The three half-filled *p* orbitals on each N atom, shown separated for clarity: (b) overlap along the *y* axis to form a *σMO:* (c) the clouds of two *πMO*s formed by overlap along the x and z axes.

atom, could overlap to form a σ *MO* with a shape somewhat like that in Fig. 4-7, containing the H 1*s* electron and the F $2p_z$ electron with spins paired to form a σ bond. (Note the *perpendicular* nodal plane.) All the other fluorine electrons may be considered *nonbonding*. Those pairs that are more external in location are called *lone pairs* and, as we shall see later, have a large influence on the shapes of molecules. The lone pairs in HF are the $2p_x$ and the $2p_y$ electrons. The 1*s* and 2*s* electrons are considered *core* electrons in this approximation.

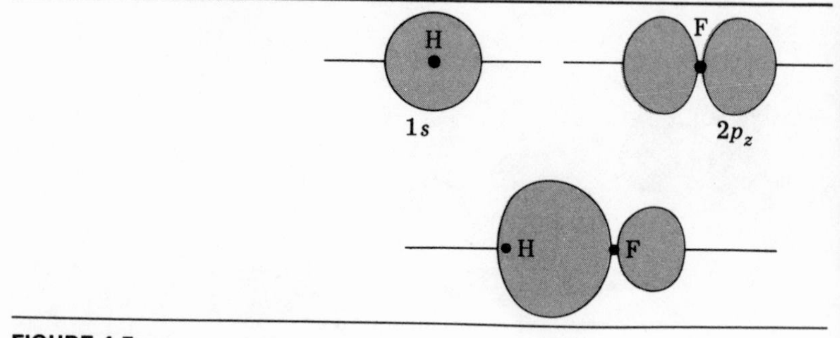

FIGURE 4-7
Formation of the HF molecule.

Suppose that we consider a larger molecule: H_2O, water. The electronic structure of the oxygen atom is

$$O\ (Z = 8): \qquad 1s^2\ 2s^2\ 2p_x^1\ 2p_y^1\ 2p_z^2$$

We shall regard the 1*s* and 2*s* electrons as core electrons. In Fig. 4-8 these are not shown, and for simplicity only one lobe of each *p* orbital is drawn. The $2p_z$ orbital (shaded) is already filled with a lone pair of electrons and is thus not available for bonding. If two hydrogen atoms with proper spin approached the p_x and p_y orbitals head on, two σ bonds, each with a shape like that in HF, would form. Accordingly we should expect the

FIGURE 4-8
The p^2 model of the water molecule.

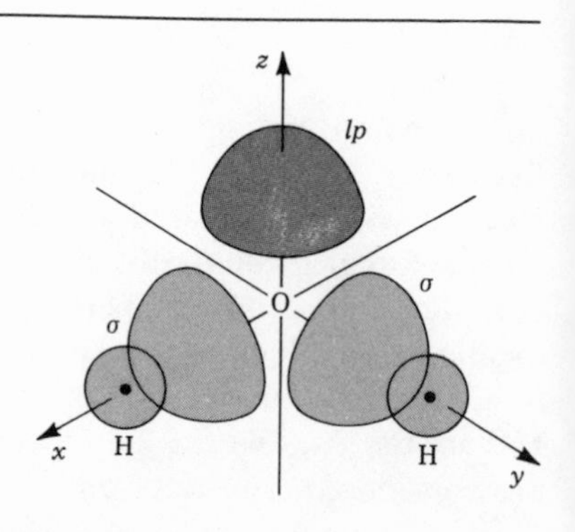

water molecule to be nonlinear, with a bond angle, determined by the nuclear positions, of about 90°, since this is the angle between the p_x and p_y orbitals. Experimental studies of H_2O indicate that the molecule is indeed bent, with a bond angle somewhat larger than our prediction, namely 105°. Nevertheless, this simple model obviously warrants further examination.

We will try it next on the ammonia molecule, NH_3. The nitrogen atom has one less electron than the oxygen atom, and thus has one electron each in the $2p_x$, $2p_y$, and $2p_z$ orbitals. As shown in Fig. 4-9, three σ bonds may form when three hydrogen atoms approach these p orbitals head on. Thus in the ammonia molecule the four atoms would not lie all in the same plane, but rather form a pyramid whose base is the three hydrogen atoms, with the N atom at the apex of the pyramid. The experimental H—N—H bond angles are 106°, considerably larger than the predicted 90°, but the general three-dimensional shape of the molecule certainly agrees with that predicted by this simple model.

In a later section, when the concept of hybridization has been introduced, we shall discuss improved models for ammonia and water molecules that predict bond angles in better agreement with experiment.

FIGURE 4-9
The p^3 model of the ammonia molecule.

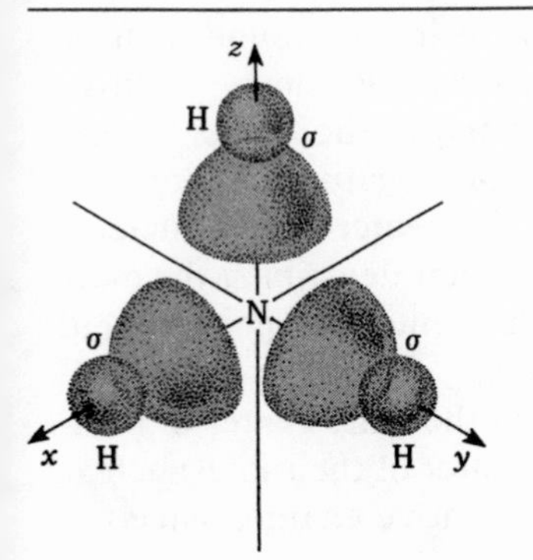

BOND POLARITY AND ELECTRIC DIPOLE MOMENTS

4-5 For homonuclear molecules like H_2, N_2, and O_2, there is no question about whether the electron pair in the bonding *MO* is shared equally by the two nuclei. Since both atoms making up the molecule are the same, both pull equally on the bonding electrons. But for heteronuclear molecules this is not true. Certain atoms of the periodic table are particularly electron-greedy (have a high *electronegativity*) and, when involved in a bond, tend to take more than their share of the bonding electron pair. The most electronegative atom is fluorine; following that are O, N, and Cl. Roughly, electronegativity decreases across the periodic table from *right to left* and down the table within a family. As the difference in electronegativities of two atoms in a heteronuclear bond increases, the *MO* cloud of the bonding pair is distorted toward the more electronegative atom, and the character of the bond departs more and more from pure covalency. In the extreme case of little or no sharing, we have an *ionic bond*, i.e., a system of a positive and a negative ion held together by electrostatic forces. This case will be discussed in detail in the next chapter.

Between these extremes, equal sharing and no sharing, we can describe bonds as being partially covalent or partially ionic, and a particularly useful vehicle for quantitatively describing these systems is the measurable molecular property called the electric dipole moment.

We can best illustrate what the dipole moment is by first considering a molecule without one, H_2. In an isolated hydrogen atom the singly negative electron in its ground state is spread about in a sphere surrounding the singly positive nucleus. The center of positive charge is the center of the nucleus, and similarly the center of negative charge (the average position of the electron) is the center of the nucleus. (Note that this is true whether the orbital describing the electron is *s*-, *p*-, or *d*-shaped.) Thus the centers of positive and negative charge *on the atom* coincide.

As two H atoms approach one another, the center of positive charge for the system is at the center of their internuclear axis, coincident with the center of negative charge, since the electron pair is equally shared. Such a molecule has no dipole moment.

For a *heteronuclear* molecule like HF, the isolated atoms still have coincident centers of negative and positive charge. As they approach, the center of positive charge is somewhere between them, closer to the heavier atom, and until the bonding pair overlaps, the center of negative charge is at the same point. If, however, when overlap occurs, the electron-greedy atom pulls the bond pair toward it, the center of negative charge will move with the pull and will no longer coincide with the positive-charge center. Molecules whose positive- and negative-charge centers do not coincide are said to possess an *electric dipole moment*, frequently represented by a little arrow or vector with tail sitting at the positive center and head pointing at the negative center.† The length of the arrow represents the magnitude of the dipole moment.

Quantitatively this dipole moment is measured in terms of a unit called a Debye (D), defined so that a positive and a negative charge (each equivalent in magnitude to the electron charge) separated by a distance of 0.1 nm has a dipole moment of 4.8 D. Generally the dipole moment μ is given by

$$\mu = zr_0$$

where z is the magnitude of the charges separated by the distance r_0.

The electric dipole moment of a molecule may be measured by placing the substance between the plates of a capacitor. In the presence of the electric field between the capacitor plates the little dipoles tend to line up (so far as their molecular motion, determined by the temperature, permits) with their positive ends pointing at the negative capacitor plate and their negative ends toward the positive plate; this orientation results in a decrease in the electric field strength between the plates and an increase in the *capacitance* of the capacitor. The capacitances of the capacitor plates when separated by a vacuum, C_0,

†Note that most physicists follow the opposite convention and point the arrow toward the positive charge.

and when separated by a substance S, C_S, are related by a factor called the *dielectric constant* ϵ of S:

$$C_S = \epsilon C_0$$

Measurement of these capacitances and consequently the dielectric constant allows one to calculate the magnitude of the dipole moment. More details are given in Ref. 4.

In the water molecule (Fig. 4-10*a*) each bond has its own dipole moment because of the electron greediness of the oxygen atom, and these moments added vectorially yield a resultant dipole moment for the molecule, represented by the dashed vector bisecting the H—O—H bond angle.

That the CO_2 molecule has a zero resultant dipole moment, even though C and O have different electronegativities, indicates that somehow the arrows representing the C—O bond dipole moments cancel. This can be true only if they are equal and pointed in opposite directions (Fig. 4-10*b*). Thus CO_2 must be a linear molecule, with the carbon atom in the center.

Obviously SO_2 (Fig. 4-10*c*), with a dipole moment of 1.61

FIGURE 4-10
Addition of dipole-moment vectors for (*a*) H_2O; (*b*) CO_2; (*c*) SO_2.

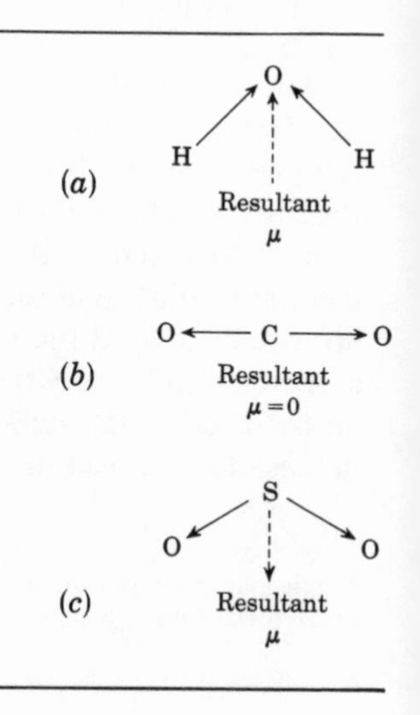

TABLE 4-1
Electric Dipole Moments of Some Common Molecules

MOLECULE	μ, D
H_2O	1.85
NH_3	1.49
CO_2	0.0
SO_2	1.61
BCl_3	0.0
CH_4	0.0

D (Table 4-1), must be bent like H_2O, so that the bond moments do not cancel.

That the dipole moments for BCl_3 and CH_4 are zero (Table 4-1) means that the geometry of these molecules must be such that the resultant of the bond dipole vectors must be zero. We shall use this information later when we discuss these molecules.

NEED FOR THE CONCEPT OF HYBRIDIZATION

4-6 We return now to a survey of the *MO* structures of simple molecules and consider $BeCl_2$, a molecule experimentally known to be linear in shape with a chlorine atom on either side of the Be atom. Like CO_2, $BeCl_2$ has no dipole moment. Now an isolated Be atom has a ground-state electronic configuration $1s^2\,2s^2$, and hence at first glance we might expect Be to behave chemically like the inert gas He, which, with its $1s$ orbital filled, shows no tendency to form molecules with other atoms. However, if we refer to Fig. 3-6, we see that for Be there is another energy level lying quite close to the outermost filled level. For He this is not true. Orbitals on the same atom which lie close to one another in energy have an unusual ability to combine with one another in an additive way, forming what are called *hybrid* orbitals. The phenomenon of hybridization is best illustrated with some pictures.

An s orbital can combine with a p orbital *on the same atom* to form two new and completely equivalent orbitals in the manner shown in Fig. 4-11. Note that the resulting unsymme-

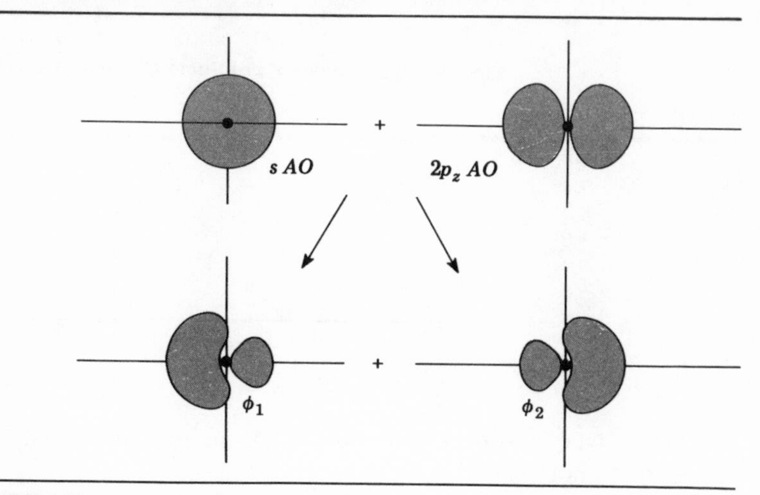

FIGURE 4-11
Formation of *sp* digonal hybrids.

trical hybrid orbitals (named ϕ_1 and ϕ_2) have properties of both the *s* and *p* orbitals which were smeared together to form them. The hybrids are "fatter" than pure *p* orbitals but have the directional characteristics of *p* orbitals. Those particular hybrids are called digonal or *sp* hybrids, since one *s* orbital and one *p* orbital were sacrificed to make them. Figure 4-12 illustrates a common abbreviation of hybrid shape.

If we assume that in the $BeCl_2$ molecule the Be atom has its two outer electrons arranged one each in two *sp* hybrids (instead of both in the 2*s*), then the molecule may be pictured as shown in Fig. 4-13, where two chlorine atoms have lined themselves up so that the two hybrids on Be are pointed directly at their half-filled $3p_z$ orbitals. Overlap creates two σ *MO*s, each with two electrons. The filled 1*s*, 2*s*, 2*p*, 3*s*, $3p_x$, and $3p_y$ orbitals on each Cl atom are not shown. The *MO* structure is consistent with the molecule's linear geometry.

At this point, and as we proceed through other examples, the skeptic may rebel at the concept of hybrid formation, proclaiming that it is only an artificial device enabling us to rationalize bonding in systems for which use of pure orbitals fails

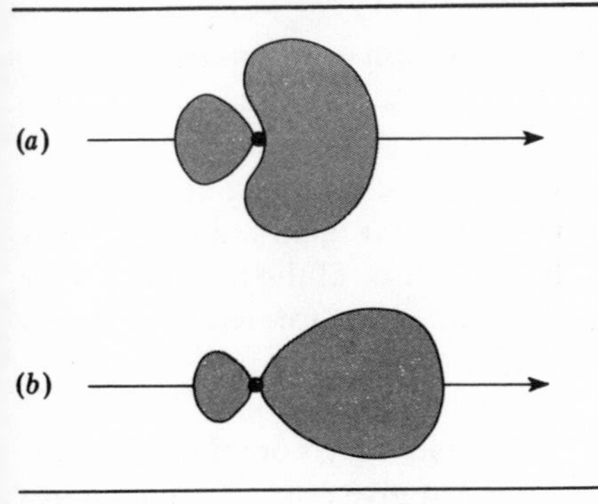

FIGURE 4-12
The actual shape of a hybrid (*a*) and its abbreviation (*b*).

to give us any reasonable picture. Then he must recall that, after all, the Schrödinger equation has been solved exactly only for the hydrogen atom and that only for one-electron species are the s, p, d, f, . . . orbitals rigorous descriptions of electron behavior. We have no right to be upset because *all* molecules cannot be described by electron pairing among pure H-like orbitals. Hybrids and *MO*s are approximations to the "true" solutions to the Schrödinger equation for molecules.

We may introduce two more common types of hybridization by examining some common carbon-containing molecules, for example, methane gas, CH_4. Experiment tells us that the four bonds in this molecule are equivalent; but when we examine the electronic structure of the isolated carbon atom, $1s^2 2s^2 2p_x^1 2p_y^1$, we see that there appear to be only two elec-

FIGURE 4-13
Structure of the linear $BeCl_2$ molecule.

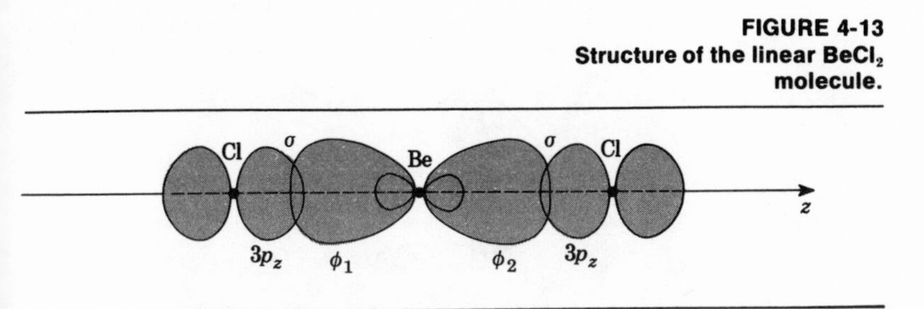

trons available for pairing and bond formation, namely, the two unpaired p electrons. Since the $2s$ orbital is quite near the $2p$ in energy (see again Fig. 3-6), we may first promote one of the $2s$ electrons to the $2p$ level, thus obtaining the "excited" configuration $1s^2\,2s^1\,2p_x^1\,2p_y^1\,2p_z^1$, which with four unpaired electrons is at least associable with four bonds. Three of the bonds, however, would be different from the fourth (that formed from the $2s$ orbital), and experiment says that this is not true.

An answer is again found in the phenomenon of hybridization. The $2s$ and the three $2p$ orbitals may combine or smear together (all four of them this time) to give four new and equivalent hybrid orbitals called sp^3 or tetrahedral hybrids. This process is pictured in Fig. 4-14. The four resulting hybrids all have shapes similar to the sp hybrids of Fig. 4-12, but here they are directed in space toward the corners of a regular tetrahedron, the carbon nucleus (and the $1s$ electron core) being at the center of the solid figure. In Fig. 4-14 the smaller lobes of the hybrids are not shown. If each of these hybrid or-

FIGURE 4-14
Formation of tetrahedral sp^3 hybrids.

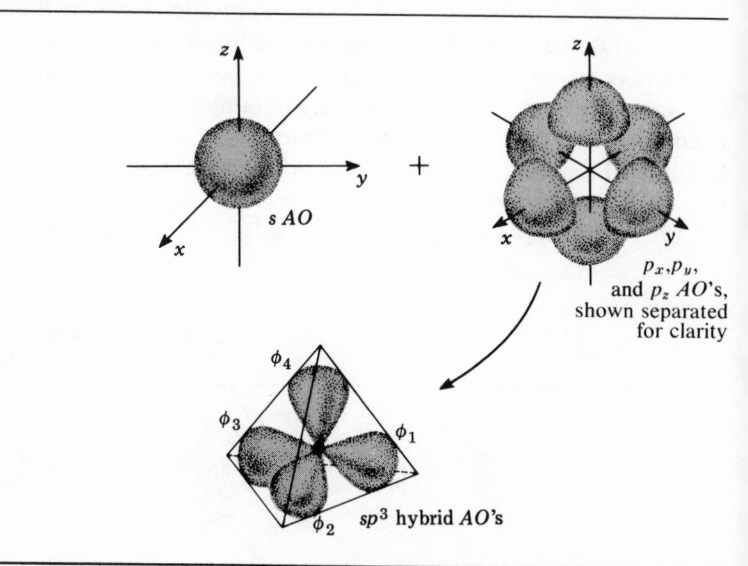

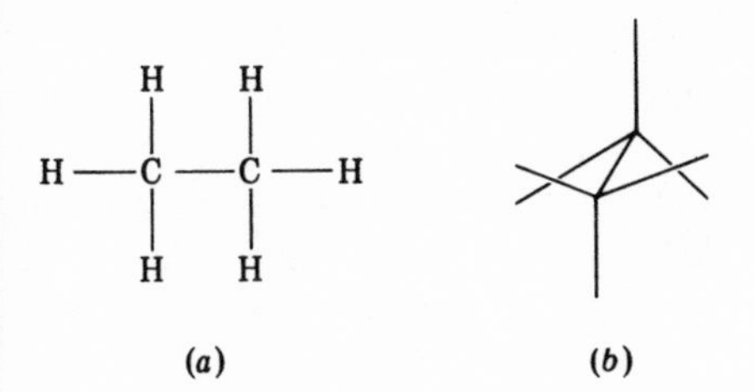

FIGURE 4-15
Structure of the ethane molecule C_2H_6. (*a*) Line formula; (*b*) stick model showing geometry and orientation of the two CH_3 groups.

bitals contains one of the four unpaired electrons of the excited carbon atom, we may suppose that bonding occurs when the four hydrogen atoms, each with one unpaired 1*s* electron, approach the four apexes of the tetrahedron and overlap with the hybrid orbitals. Four equivalent σ bonds form. The shape of the CH_4 molecule would consequently be like that of a tetrahedron; experimentally this is true. Furthermore, as stated in Sec. 4-5, the overall electric dipole moment of CH_4 is zero, a fact consistent with its highly symmetrical shape. The resultant of the four vectors representing bond dipole moments is zero. (A student familiar with vector analysis may easily prove this. See Exercise 14.)

Now let us look at the somewhat more complex molecule ethane, C_2H_6. A line formula showing how the atoms are connected is given in Fig. 4-15*a*. The molecule, however, is not planar, but instead is three-dimensional, with bonds arranged as shown in the stick model (Fig. 4-15*b*). As in CH_4, the four bonds around each carbon atom point toward apexes of a tetrahedron; three of these bonds are to hydrogen atoms, the fourth to the adjacent carbon atom. In the most stable conformation of the molecule the front CH_3 group is rotated by 60° with respect to the back CH_3 group. With this experimental information, it is easy to explain what the bonding electrons are doing. Each carbon atom is in an sp^3 state of hybridization; a σ bond is formed between the carbons by overlap of two hybrid orbitals along the carbon-carbon internuclear axis; the six hydrogen atoms are bonded in σ fashion with the remaining six tetrahedral hybrids. It is known experimentally that the two CH_3 groups may rotate with respect to one another about the σ bond connecting them. A relatively small amount of energy is necessary to overcome the repulsion of bond pairs and hydrogen nuclei between the two groups.

Proceeding now to a related molecule, ethylene, C_2H_4, we first examine the experimental facts. The atoms of ethylene, unlike those of ethane, all lie in the same plane, i.e., are coplanar. The total strength of the bond(s) connecting the two carbon atoms is considerably greater than that in ethane (almost, but not quite, twice as much), and, furthermore, the two CH_2 groups *cannot* rotate with respect to one another. These latter two facts indicate the existence of more than just a σ bond between the carbon atoms; i.e., some π formation is likely. (The student should satisfy himself with a model that two CH_2 molecular fragments connected by just a σ bond may rotate with respect to one another *without breaking the bond* but that two fragments connected by a σ and a π bond are restricted in rotation unless the π bond is broken.)

In order to rationalize the structure of this molecule, let us again consider the promoted configuration of the carbon atom, $1s^2\,2s^1\,2p_x^1\,2p_y^1\,2p_z^1$. This time, instead of forming sp^3 hybrids, we shall leave one of the p electrons (arbitrarily the $2p_z$) in its pure orbital and mix or hybridize the $2s$ with the $2p_x$ and the $2p_y$ orbitals. Mixing these three will give us three equivalent hybrid orbitals, all lying in the xy plane, since we have mixed into the hybrids only the x- and y-preferring p orbitals. The shapes and orientation of the hybrids, called trigonal or sp^2 hybrids, are illustrated in Fig. 4-16. Note the 120° bond angles.

After forming two sets of these trigonal hybrids, one set on each carbon atom, we shall line up the two carbons so that all the hybrids are lying in the same plane and so that one on each carbon atom overlaps to form a carbon-carbon σ bond (Fig. 4-17). The other two hybrids on each carbon form σ bonds with hydrogen atoms. The resulting planar structure is called the *σ framework* of the molecule; it lies in the xy plane.

The one electron on each carbon atom which we left in a

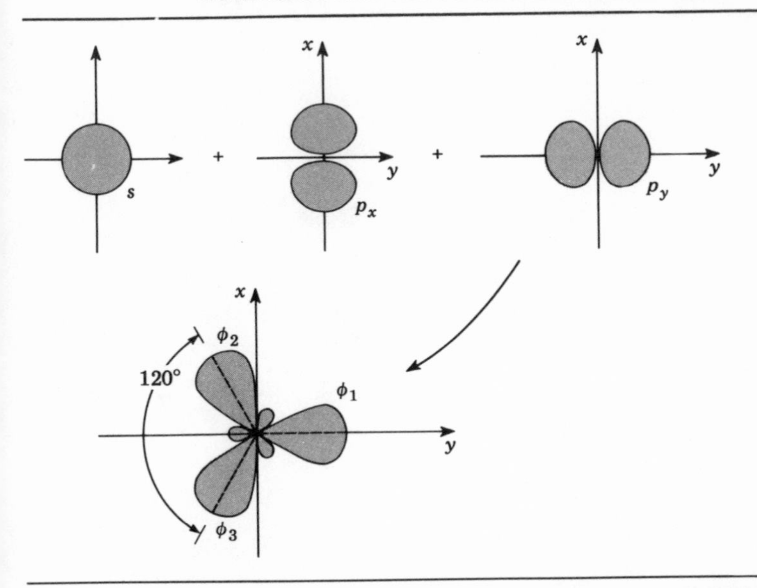

FIGURE 4-16
Formation of sp^2 trigonal hybrids.

pure p_z unhybridized state will occupy an orbital perpendicular to the plane of the σ framework (Fig. 4-18*a*). Overlap of these two p_z orbitals sideways gives rise to a π *MO*, illustrated in

FIGURE 4-17
The σ framework of ethylene, C_2H_4.

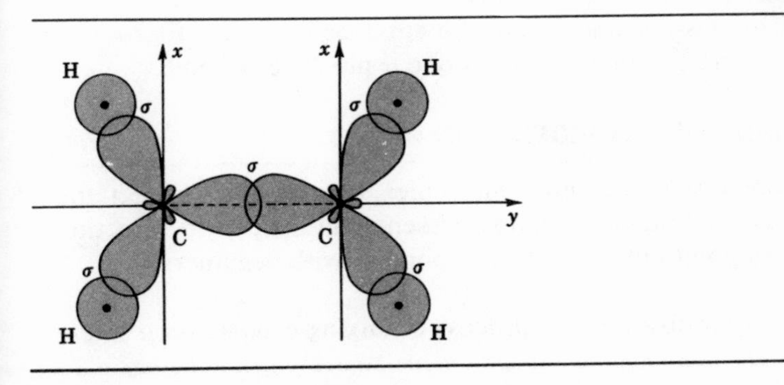

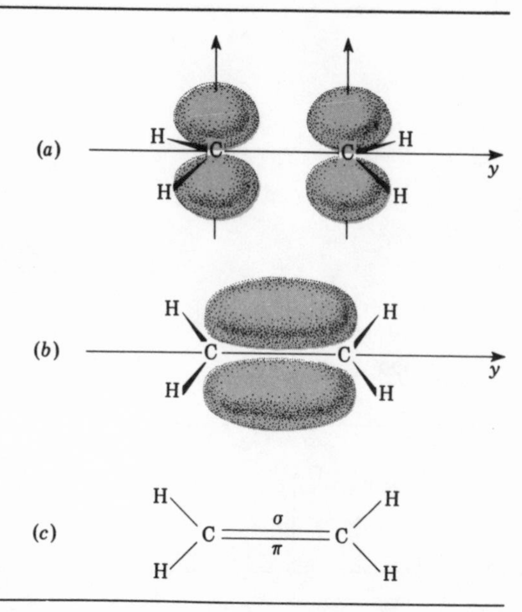

FIGURE 4-18
Pi molecular orbital formation in ethylene. (*a*) The p_z orbitals before overlap; (*b*) the π cloud after overlap; (*c*) the line formula indicating σ and π bonds.

Fig. 4-18*b*. Since the energy of a π bond is less than that of a σ bond, the amount of energy necessary to rip ethylene into two CH_2 fragments is less than twice the energy for fragmentizing ethane into two CH_3 groups.

The classical line formula for ethylene is given in Fig. 4-18*c*, where the double line means one σ plus one π bond.

A SUMMARY OF HYBRIDIZATION RULES

4-7 Before discussing more molecules we pause to collect the rules of hybrid construction presented in the previous section and to point out some pitfalls popular with beginners.

1 Hybridization is a process of mixing orbitals *on a single* atom (or ion). In contrast, molecular orbital formation requires mixing of orbitals originally centered on *different* atoms.

2 Only orbitals of similar energies can be mixed to form good hybrids. For our purpose this will mean that the orbitals usually must belong to the same "energy group" as shown in Fig. 3-6.

3 The number of orbitals mixed together always equals the number of hybrids obtained.

4 In hybridization we mix a certain number of *orbitals*, not a number of electrons. For example, if we wish to form digonal (sp) hybrids on an atom in a three-electron configuration $2s^2\,2p^1$, we mix the s and p orbitals to get two hybrids ϕ_1 and ϕ_2, and then arrange the electrons among them, possibly as $\phi_1^2\phi_2^1$.

5 Once an orbital has been used to build a hybrid it is no longer available to hold electrons in its "pure" form. For example, in item 4 above, a configuration $2s^1\,\phi_1^1\,\phi_2^1$ is illegal, for the $2s$ orbital no longer exists as such.

6 Most hybrids are similar, but they are not necessarily identical in shape; they differ from one another largely in orientation in space. A proper representation of the sp hybrid is shown in cross section in Fig. 4-12. The three-dimensional shape is cut out of space by rotating the z axis without translating it in space, i.e., rolling it between the fingers.

7 Since s orbitals are nondirectional in xyz space, they add no direction when contributing to hybrids. They add "plumpness" only.

8 Other orbitals with pronounced directions in space (p_x, p_z, d_{xy}, d_{yz}, etc.) determine the directional properties of hybrids. Mixing only x-liking and y-liking orbitals with s orbitals gives hybrids preferring the xy plane; mixing an x-liking orbital with an s orbital yields hybrids liking the x axis.

9 For equivalent hybrids, the orientation in space is determined by (a) the number of orbitals mixed and consequently the number of hybrids obtained, (b) which of the x, y, and z directions are preferred by the orbitals when "pure," and (*c*) the assumption that the electrons which will occupy the hybrids will try to avoid one another as far as possible within limitation *b*. For example, suppose we wish to form three equivalent hybrids by mixing the $2s$, $2p_x$, and $2p_y$ orbitals on an atom. Condition *b* demands that their predominant directions lie in the *xy* plane; condition *c* divides the 360° of the *xy* plane into three parts, making the angle between hybrids 120°.

10 The particular type of hybrid chosen for a structure discussion is determined by the experimentally known geometry of the molecule. (However, if it is not known, we still may make educated guesses about its shape by comparing it to related molecules.) Bond angles of 120° should hint of sp^2 hybrids, linear systems of *sp* hybrids, and tetrahedral shapes or 109° bond angles of sp^3 hybrids.

THE NEON MOLECULAR ISOELECTRONIC SERIES

4-8 The noble-gas atom Ne and the molecules HF, H_2O, NH_3, and CH_4 all have the same number of electrons (are isoelectronic), and all have the same *total* number of positive nuclear charges. The central atoms all have related external electronic structures; in going from C to Ne we simply fill the $2p$ level with its quota of electrons. Thus it is not surprising that the structures of these four molecules are similar. To illustrate this and to provide simultaneously the "better" picture of H_2O and NH_3 promised in Sec. 4-4, let us start with the tetrahedral picture of CH_4, a symmetrical molecule with four bond pairs separated by 109° bond angles (Fig. 4-19).

Imagine that we pluck one of the H nuclei (an H^+) from a σ bond in CH_4 and merge it with the carbon atom nucleus, increasing *Z*, the atomic number of carbon, by 1 and thus converting it to a nitrogen nucleus. The former bond pair is now a lone pair and overall what we now have is an NH_3 molecule (Fig. 4-19) still with the experimentally known pyramidal shape (determined by the nuclei) but now with bond angles of 109°. Thus the bond angle predicted by an sp^3 model of NH_3 is much closer to the experimentally known 106° than the p^3 model of Sec. 4-4 with its 90° angles.

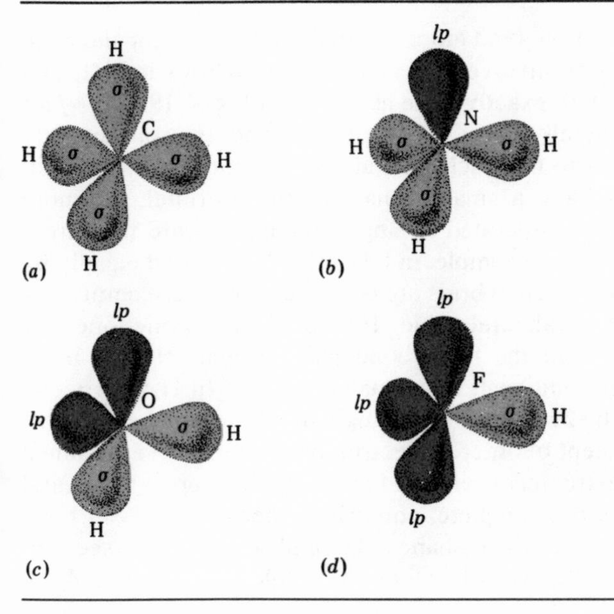

FIGURE 4-19
***sp*^{3} models of CH_4, NH_3, H_2O, and HF.**

Pushing another proton into the central nucleus gives us a water molecule with two σ bond pairs separated by 109° and two lone pairs completing the tetrahedron. One more proton push gives us HF, with three lone pairs and one σ bond.

Rationalizing the 109° perfect sp^3 bond angles down to 106° for NH_3 and 105° for H_2O is easy if we accept a rule ordering the strengths of pair-pair repulsions:

$$lp\text{-}lp > lp\text{-}bp > bp\text{-}bp$$

meaning that repulsion between lone pairs is greatest, repulsion between bond pairs the least. In CH_4 we have all *bp-bp* repulsions and a symmetrical molecule. In NH_3 we introduce one *lp*, and since *lp-bp* repulsion is greater than *bp-bp* repulsion, the *bp*'s move together to escape the *lp*, thus decreasing the bond angle to 106°. Introducing another *lp*, in H_2O, decreases the angle further, to 105°.

An alternative (and more correct) explanation for the bond angle variation involves the fact that the hybrids in NH_3 and H_2O are not all exactly alike as shown in Fig. 4-19. *Lone-pair* orbitals generally "hog" the $2s$ orbital and are consequently fatter, closer to the nucleus, and less directional. As a result, *bond pairs* have a smaller share of the s orbital, are more p-like, and are separated by angles tending toward the pure p angle of 90°. For example, in CH_4, the $2s$ is shared equally by four bond pairs, all hybrids are equivalent and are separated by the perfect tetrahedral angle, 109°28′. In NH_3 one lone pair hogs the $2s$ and the three bond pairs become slightly more p-like, with bond angles decreasing to 106°. In H_2O two lone pairs hog the $2s$ and the bond angle decreases further to 105°.

The concept of ordered electron pair repulsions, as outlined above, is extremely useful and has been developed by several scientists into a complete, non-orbital theory capable of powerful predictions about shapes of complex molecules (see Ref. 16, 17, and 18). We will call on these rules again in Sec. 4-13.

OTHER SIMPLE MOLECULES

4-9 Before introducing more new concepts, we shall look at two or three examples of how to build molecular orbitals, emphasizing in this section "method of attack."

Consider the molecule H_2CO, formaldehyde, with the nuclear framework illustrated in Fig. 4-20*a*. Experiment tells us that this molecule is planar and that the angle between the two hydrogen atoms is slightly less than 120° (*Hint: sp^2* hybrids on carbon!). With this knowledge in hand we can proceed to promote and hybridize the carbon to the $\phi_1^1\phi_2^1\phi_3^1 2p_z^1$ state, where the ϕ's are sp^2 hybrids in the xy plane, the plane of the molecule. We orient the hybrids toward the O and H atoms, as in Fig. 4-20*b*. The oxygen atom has the configuration $1s^2 2s^2 2p_x^2 2p_y^1 2p_z^1$, and through overlap of its $2p_y$ orbital with ϕ_3 of carbon a σ bond forms. The lone pair $2p_x$ on oxygen lies in the plane of the σ framework. ϕ_1 and ϕ_2 form σ bonds with the hydrogens to complete the σ framework. Perpendicular to this plane, the $2p_z$ orbitals on carbon and oxygen overlap sideways to form a π bond (Fig. 4-20*c*). The line

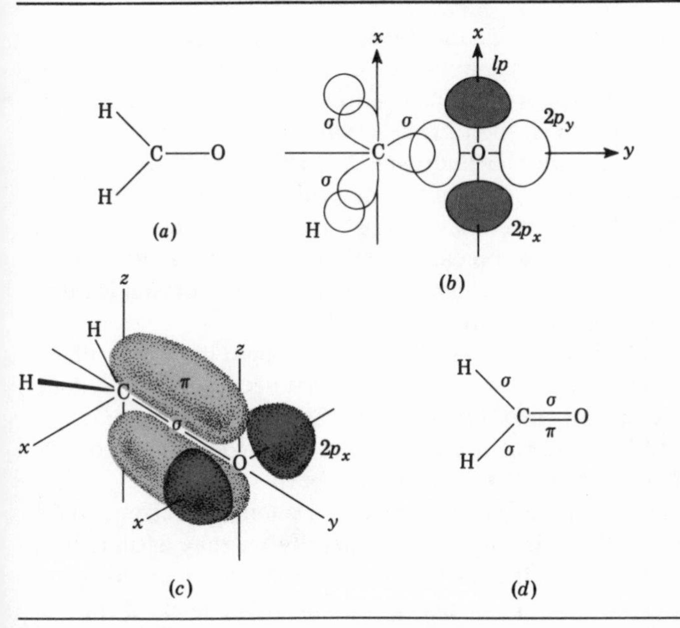

FIGURE 4-20
Structure of the formaldehyde molecule. (*a*) Position of the nuclei; (*b*) the σ structure; (*c*) the π structure; (*d*) line formula showing σ and π bonds.

formula with proper identification of bonds is shown in Fig. 4-20*d*, and the molecule is "finished."

A second method of attack, which will frequently prove useful when an electron-pairing scheme for a molecule is not immediately obvious, is the "reservoir" approach. Again consider formaldehyde—its shape and the hint of sp^2 hybridization. The external structures of atoms involved are

$$\begin{aligned} &\text{C:} \quad 1s^2\,2s^2\,2p^2 \\ &\text{O:} \quad 1s^2\,2s^2\,2p^4 \\ &\text{H:} \quad 1s^1 \quad \text{(two of these)} \end{aligned}$$

Let us strip off all external electrons from these atoms (4 from carbon, 6 from oxygen, and 2 from the hydrogens) and put the 12 electrons in an imaginary electron reservoir or bucket while we deal with empty orbitals. Once again we construct sp^2 hybrids from the carbon $2s$, $2p_x$, and $2p_y$ orbitals and overlap two of them with the hydrogen $1s$ orbitals to form σ *MO*s (empty). We may construct *sp* hybrids on the oxygen atom (from the $2s$ and the $2p_y$ orbitals) and overlap one with the remaining sp^2 hybrid on carbon to form another empty σ *MO*. The empty σ framework (which includes every orbital located in the *xy* plane) is shown in Fig. 4-21.

(Actually in our previous method we could have hybridized the oxygen atom as we did here. It is not necessary, but it does have the advantage of removing the $2s$ oxygen electrons away from the bonding region, converting them to a lone pair. sp^2 hybrids on oxygen would also have been suitable.)

We now remove from our reservoir enough electrons to *fill completely* the σ framework. (Generally we may assume that positions in the σ framework are more stable or are lower in energy than positions in the π structure.) We place 2 each in the C—H sigma bonds, 2 in the C—O sigma bond, 2 in the *lp* $2p_x$ on oxygen, and 2 in the *lp sp* hybrid on oxygen. A total of 10 electrons are gone; 2 are left in the bucket, and only 2 orbitals are left unused, $2p_z$ on carbon and $2p_z$ on oxygen. We

FIGURE 4-21
The σ framework of formaldehyde with hybridized oxygen.

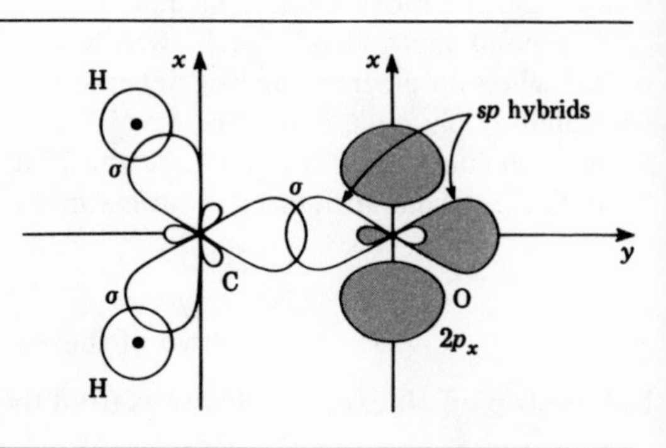

overlap these to form a π *MO* and fill it with the last 2 electrons. The result is essentially the same as before (Fig. 4-20*d*).

Both the "pairing" and the "reservoir" methods have their advantages. We recommend mastery of both.

DELOCALIZED π MOLECULAR ORBITALS

4-10 In all the molecules discussed so far the π *MO*s were *localized* between two atoms, for example, between the two carbon atoms in ethylene in Fig. 4-17. Frequently, however, we encounter systems for which it is difficult to combine orbitals in the electron-pairing approach so that a unique set of π *MO*s results. Then we must introduce the concept of *resonance* or delocalized π bonds. These ideas are easily illustrated with the SO_2 molecule.

SO_2 is nonlinear (recall that its dipole moment was not zero), with the nuclear framework suggested in Fig. 4-22*a*. The configurations of the isolated atoms are

$$\text{S}: \quad 1s^2\,2s^2\,2p^6\,3s^2\,3p^4$$
$$\text{O}: \quad 1s^2\,2s^2\,2p^4 \quad \text{(two of these)}$$

We remove all electrons on sulfur with $n = 3$ and all on each oxygen with $n = 2$, placing a total of 18 in our reservoir and leaving all others as core electrons.

Since the nuclear framework hints of sp^2 hybridization, we construct these hybrids from the s, p_x, and p_y orbitals on each atom, overlap them in the xy plane, as shown in Fig. 4-22*b*, and assign 14 of our 18 electrons to the σ framework, 2 each in the two σ *MO*s and 10 as the lone pairs. Our 4 remaining electrons must be accommodated in the π framework built from the p_z orbitals perpendicular to the σ framework (Fig. 4-22*c*). But we find that we can do this in two entirely equivalent ways (Fig. 4-22*c* and *d*), with an *lp* on the right and a π *MO* on the left, or vice versa. Neither structure alone is adequate, for each predicts a molecule with one S—O linkage different from the other. Experiment says that SO_2 is symmetrical. If we want to retain the electron-pair picture, we must say that the true description of the molecule is something between these two pictures and denote this with a double-headed arrow placed

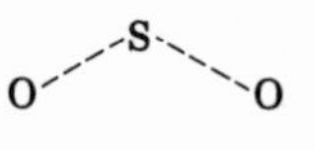

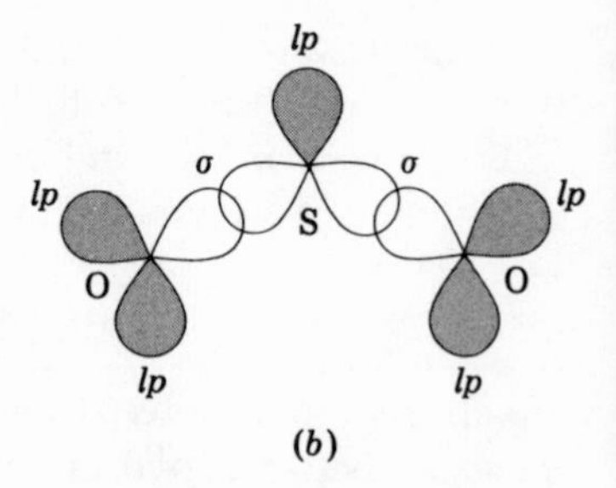

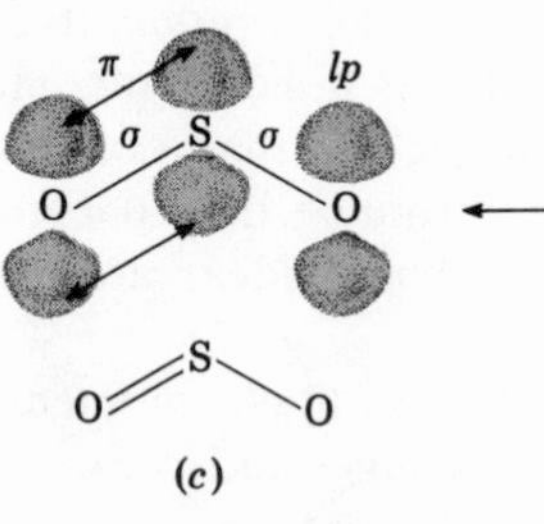

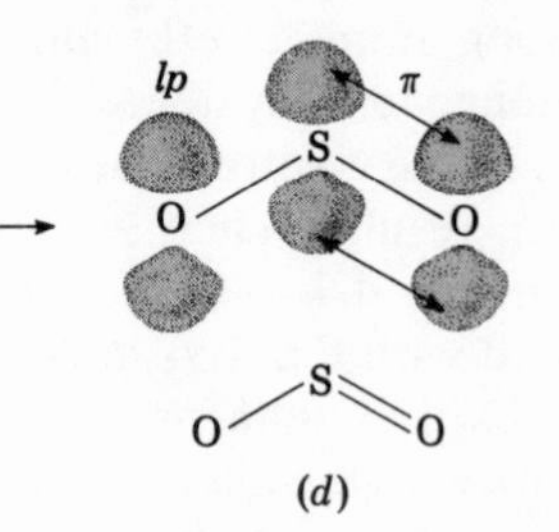

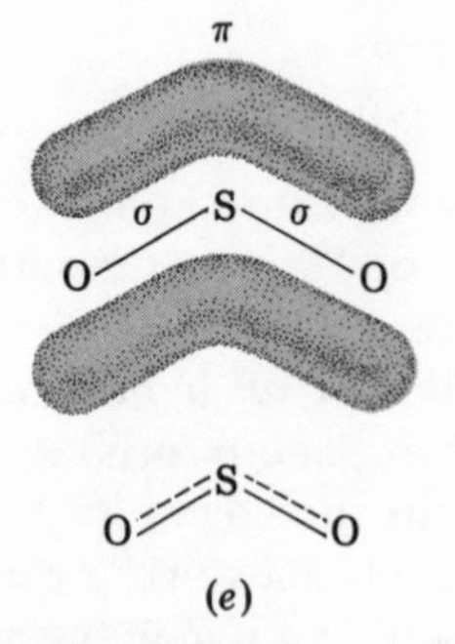

FIGURE 4-22
Structure of the SO_2 molecule. (*a*) Position of the nuclei; (*b*) σ framework in the *xy* plane; (*c,d*) the two resonance structures with localized π bonds; (*e*) the delocalized π cloud. In (*c*), (*d*), and (*e*) the classical line formulas are shown below the orbital pictures.

between them. This particular type of arrow is associated with the term *resonance* and means *not* that the double bond flips from side to side but rather that the two forms are the best approximations that we can make to the true structure within the electron-pair model. A better picture for the π structure is shown in Fig. 4-22*e*, which can be regarded as a superposition of structures *c* and *d*. Here the π cloud appears *delocalized*, spread over three centers, and in the cloud (which is not a single *MO* but a smear of *MO*s and atomic orbitals) there are wandering four π electrons. The dotted line in the structural formula for SO_2 is commonly used in textbooks as a symbol for delocalized π bonds.

As a second example of delocalization we consider the molecule benzene, C_6H_6, which according to experiment is a planar molecule, with all 12 atoms in the same plane. The six carbon atoms are at the corners of a regular hexagon (Fig. 4-23*a*), each bonded to a single H atom located outside the hexagon. Assumption of sp^2 hybridization of each carbon easily leads to a σ framework containing 24 electrons (Fig. 4-23*b*).

On each carbon atom in the σ framework there remains one electron in a pure *p* orbital perpendicular to the trigonal hybrids and thus to the σ framework (Fig. 4-23*c*). Sideways overlap into localized π *MO*s can occur in two ways, illustrated with the line drawings in Fig. 4-23*d*. Since all the carbon-carbon bonds in benzene are equivalent, neither structure alone can be correct and we must employ the resonance arrow. Alternatively, Fig. 4-23*e* shows the hexagonal, doughnut-shaped clouds, above and below the σ framework, obtained when the two localized π *MO* structures are superimposed. The doughnuts accommodate the six π electrons (just as the three π *MO*s held a total of 6 electrons), and in these clouds the π electrons wander around, belonging now to no particular carbon atom and moreover to no particular *pair* of carbon atoms. The π electrons are "delocalized." Recognition of delocalization in benzene has led many authors of organic chemistry textbooks to adopt the "dotted-circle" representation (Fig. 4-23*f*) in preference to diagrams showing three localized π bonds.

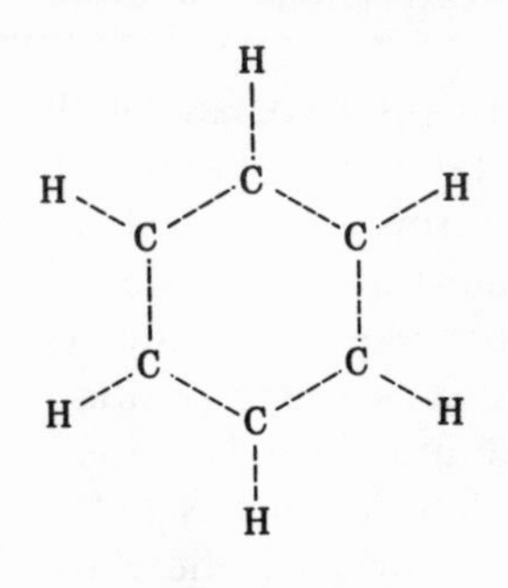

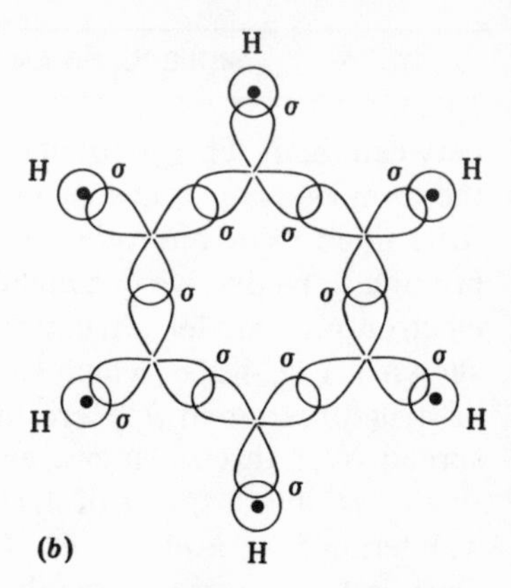

FIGURE 4-23
The structure of the benzene molecule, C_6H_6. (*a*) Position of the nuclei; (*b*) the σ framework; (*c,d*) the two resonance structures showing localized π bonds; (*e*) the delocalized π colud; (*f*) line formula indicating delocalized π electrons by a dotted circle.

A MORE DETAILED TREATMENT OF MOLECULAR ORBITALS IN DIATOMIC MOLECULES

4-11

Earlier, when we presented a naïve description of molecular orbitals in homonuclear molecules (H_2, N_2), based largely on electron-pairing concepts, we purposely avoided application to "problem" molecules, which cannot be treated satisfactorily within the simple model. In this section we shall outline a more rigorous approach to *MO*s, which can account for such problems as (1) O_2 and its paramagnetism† and (2) He_2 and its nonexistence under normal conditions.

We examine the oxygen molecule first. Experiment indicates that O_2 is essentially double-bonded and exhibits paramagnetism due to two unpaired electrons. In the simple approach we should start by considering isolated oxygen atoms with configurations

$$\text{O:} \qquad 1s^2\,2s^2\,2p_x^2\,2p_y^1\,2p_z^1$$

according to Hund's rule. Taking the z axis as the internuclear axis, we form a σ *MO* by overlapping the two $2p_z$ orbitals and, on appropriate orientation of the $2p_y$ orbitals, form a π *MO*. Occupation of the σ and π *MO*s corresponds to a normal double bond. However, we now have no unpaired electrons left to explain the paramagnetism. Retaining the double bond, we could resort to some rather artificial devices to obtain unpaired electrons, for example, exciting one electron from a $2p_x$ lone pair to a higher state; but the symmetry of the molecule argues against this. We could unpair the two $2p_x$ *lp*'s, excite one electron from each into higher states, and then pair the remaining $2p_x$ ones in a π *MO*, but this approach seems to have little logic recommending it within the simple electron-pair approach.

In search of a more reasonable explanation let us recall the principles of hybridization. When mixing orbitals *on one atom* to form hybrids, we always conserved the *number* of orbitals;

†Paramagnetism is the phenomenon associated with the magnetic behavior of molecules or solids containing unpaired electrons. In the presence of a magnetic field, attractive forces are exerted on the substance, and measurement of these (Chap. 6) enables us to estimate the number of unpaired electrons per molecule.

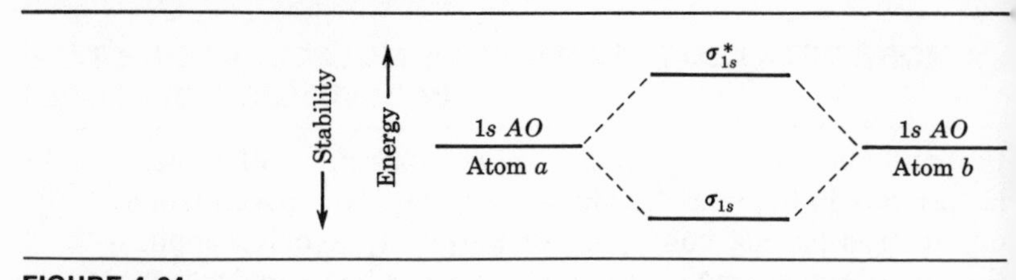

FIGURE 4-24
Formation of bonding and antibonding *MO*s.

for example, mixing four pure orbitals gave four hybrid orbitals, etc. In a sense we always conserved *allowed places for electrons to live*. A similar rule holds for the mixing of orbitals on *different* atoms to form *MO*s. Let us reconsider the H_2 molecule with this in mind. Two $1s$ orbitals merge as the molecule forms to produce a σ_{1s} *MO* of greater stability than either of the $1s$ orbitals, and *also*, although we have neglected it thus far, a less-stable *MO* called the σ^*_{1s}. The relative energies of these are shown in Fig. 4-24.

The boundary diagram for the σ_{1s} *MO* has already been discussed (Fig. 4-25) and in three dimensions is rather ellipsoidal in shape.

The σ^*_{1s} *MO* has a boundary diagram possessing a nodal plane perpendicular to the z axis between the nuclei (Fig. 4-25), and the electron cloud is shifted away from the bonding region. Note that the *MO* is still σ in nature, for the nodal plane does not *contain* the internuclear axis. In three dimensions the *MO* is dumbbell-shaped. Because of its higher energy, the σ^*_{1s} is called an *antibonding MO* (*ABMO*) to distinguish it from the bonding *MO* (*BMO*) σ_{1s}. Antibonding *MO*s are usually marked with asterisks.

In general, *for homonuclear diatomic molecules* any two like atomic orbitals can merge to form two *MO*s, one of a bonding type more stable than the isolated orbitals and another of an antibonding type less stable than the isolated orbitals. For molecules larger than H_2, where the z direction is assumed to be the internuclear axis between atoms a and b, *these* atomic orbitals unite to form *these* molecular orbitals

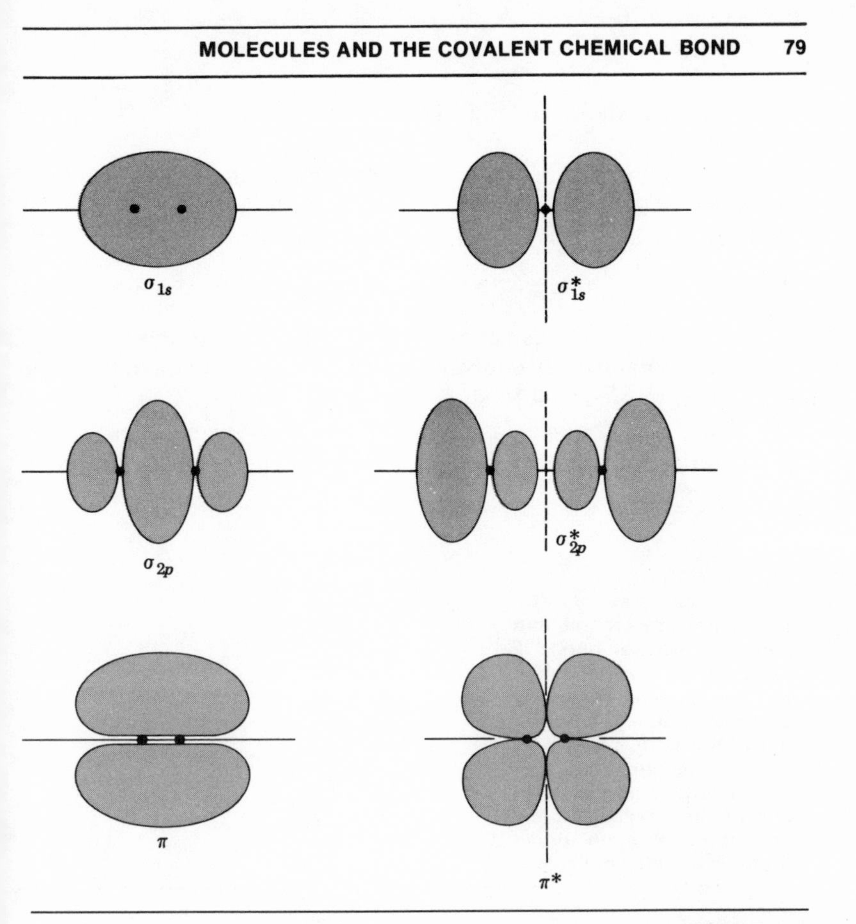

FIGURE 4-25
Contour diagrams for bonding and antibonding σ_{1s}, σ_{2p}, and π molecular orbitals.

$2s_a$ and $2s_b$	σ_{2s} and σ^*_{2s}
$2p_{za}$ and $2p_{zb}$	σ_{2p} and σ^*_{2p}
$2p_{ya}$ and $2p_{yb}$	π_y and π^*_y
$2p_{xa}$ and $2p_{xb}$	π_x and π^*_x
$3s_a$ and $3s_b$	σ_{3s} and σ^*_{3s}

etc.

The shapes of the *BMO*s have been discussed previously

and are shown in Fig. 4-25 along with the *ABMO* contour diagrams.

For most homonuclear diatomic molecules built of atoms of period 2 elements, an approximate ordering of the molecular energy levels is, according to experiment,

$$\sigma_{1s} < \sigma_{1s}^* < \sigma_{2s} < \sigma_{2s}^* < \pi_x = \pi_y < \sigma_{2p} < \pi_x^* = \pi_y^* < \sigma_{2p}^*$$

π_x and π_y orbitals have the same symmetry around the z axis, are geometrically equivalent, and consequently have the same energy. Figure 4-26 shows these *MO* energies displayed in a diagram. The spacings between the levels were chosen arbitrarily and are not meant to represent experimental energy differences.

FIGURE 4-26
Relative energies of *MO*s for diatomic molecules and molecule ions of period 2 atoms up to and including N_2. For O_2 and F_2, σ_{2p} is below π_x and π_y. Occupation of these *MO*s for the N_2 molecule is illustrated. Relative energies and occupation of the original atomic orbitals on nitrogen atoms A and B are also indicated.

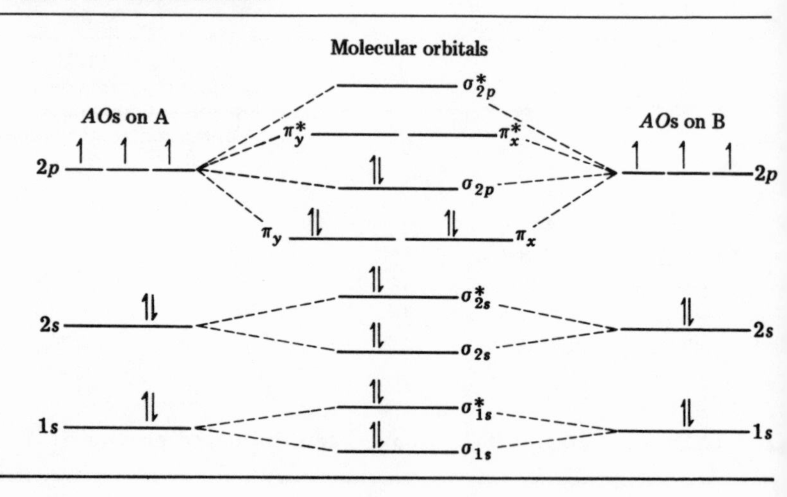

Configurations of molecules can be found by applying the *Aufbau* principle to this array of *MO*s. For O_2 (16 electrons) we have

$$O_2: \quad \sigma_{1s}^2 \sigma_{1s}^{*2} \; \sigma_{2s}^2 \sigma_{2s}^{*2} \pi_y^2 \pi_x^2 \sigma_{2p}^2 \pi_y^{*1} \pi_x^{*1}$$

The last two electrons enter the degenerate π_y^*-π_x^* level with spins unpaired according to Hund's rule, and thus the paramagnetism of O_2 is explained. To account for its double bond, we define a quantity called bond order (*BO*), such that *BO*s of 0, 1, 2, and 3 correspond to classical no-bond, single, double, and triple bonds.

$$BO = \tfrac{1}{2}(\text{no. electrons in } BMO\text{s} - \text{no. electrons in } ABMO\text{s})$$

For O_2 there are 10 bonding electrons (with no asterisk) and 6 antibonding electrons (with asterisk); thus $BO = 2$, which corresponds to a double bond.

For N_2 (which would have the same configuration as O_2 minus the two highest-energy electrons) the number of *BMO* electrons is 10, of *ABMO* electrons 4; thus the bond order is 3, and the bond is a triple one. This we had already decided, on the basis of the simpler theory.

We can now discuss the instability of He_2. This molecule (4 electrons) would have the configuration $\sigma_{1s}^2 \sigma_{1s}^{*2}$ with 2 bonding and 2 antibonding electrons, a *BO* of zero, corresponding to no bonds, no molecule. On the other hand the molecule-ion He_2^+ would have the configuration $\sigma_{1s}^2 \sigma_{1s}^{*1}$ and a net *BO* of $\frac{1}{2}$, corresponding to a "half-bond" or a one-electron bond. He_2^+ ions are known experimentally.

A rather classical problem very neatly explained by the *MO* theory concerns the dissociation energies of O_2 and its molecule-ion O_2^+ relative to the dissociation energies of N_2 and N_2^+:

PROCESS	ENERGY REQUIRED (kJ/mol)
$N_2 \rightarrow 2N$	941
$N_2^+ \rightarrow N + N^+$	842
$O_2 \rightarrow 2O$	493
$O_2^+ \rightarrow O + O^+$	637

$D(O_2)$ is smaller than that of its ion, yet $D(N_2)$ is larger than that of its ion. If we compute bond orders for these species and recognize that the larger the BO, the more energy it will take to break the molecule in two, we see that O_2, with $BO = 2$, *should* be dissociated with less energy than O_2^+ ($BO = 2\frac{1}{2}$). However, in going from N_2 ($BO = 3$) to its ion N_2^+, we remove an electron from a BMO, so that the ion BO is only $2\frac{1}{2}$. Thus $D(N_2) > D(N_2^+)$.

Since higher bond orders are associated with shorter bond distances, we can use computed BOs for a molecule and its ions to predict relative bond distances; for example, see the trend in oxygen and ions below:

MOLECULE	O_2^+	O_2	O_2^-	O_2^{2-}
BO	$2\frac{1}{2}$	2	$1\frac{1}{2}$	1
R_e (nm)	0.112	0.121	0.126	0.149

However, bond orders cannot be used to predict relative bond lengths of different molecules. Li_2 ($R_e = 0.267$ nm) and F_2 ($R_e = 0.142$ nm) both have $BO = 1$! As a general rule, single bond lengths decrease across a row in the periodic table because atom size decreases.

With caution one may extend the MO diagram of Fig. 4-26 to molecules of atoms with $3s$ and $3p$ electrons by adding the sequence

$$\sigma_{3s} < \sigma_{3s}^* < \pi_{3p_x} = \pi_{3p_y} < \sigma_{3p} < \pi_{3p_x}^* < \pi_{3p_y}^* < \sigma_{3p}^*$$

P_2, like N_2, is thus predicted to have a triple bond, and S_2, like O_2, a double bond. Again, however, these qualitative bond orders cannot be used to predict relative bond strengths or bond distances of molecules in different rows of the periodic table. P_2 and N_2 both have $BO = 3$, but $D(P_2) = 477$ kJ/mol, $R_e = 0.189$ nm and $D(N_2) = 941$ kJ/mol, $R_e = 0.110$ nm. As a general rule, bond strengths decrease and bond distances increase as we go down a column in the periodic table, the reason being larger, more diffuse $n = 3$ bonding orbitals and consequent thinner electron glue between the atoms.

The MO energy sequence of Fig. 4-26 with the $n = 3$ extension described above may also be applied to heteronuclear diatomic molecules and molecule-ions provided that the atoms are near neighbors in a row in the periodic table, for example, NO, OF, CN^-, PS, and even CO and NF (but not LiF, MgO, BeS). As a sample, the paramagnetic molecule NO has a bond order of $\frac{5}{2}$ (a double bond plus a 1-electron bond) according to the configuration

$$\sigma_{1s}^2 \sigma_{1s}^{*2} \sigma_{2s}^2 \sigma_{2s}^{*2} \pi_x^2 \pi_y^2 \sigma_{2p}^2 \pi_x^{*1}$$

In general, diatomic *MO* theory contradicts none of our previous conclusions and permits us to make very useful predictions about molecules for which the simpler electron-pair approach is inadequate.

HYBRIDS INVOLVING *d* ORBITALS

4-12 Three other types of hybrids are frequently useful in discussions of bonding in molecules containing atoms with *d* electrons or with low-lying empty *d* orbitals. We shall consider their properties briefly here and illustrate their use in Sec. 4-13.

Four dsp^2 hybrids of *square planar* symmetry may be constructed by mixing the $d_{x^2-y^2}$, s, p_x, and p_y orbitals. The four hybrids lie in the xy plane with neighboring lobes separated by 90°, as shown in Fig. 4-27. Each lobe represents one hybrid, and each may be used to form one σ bond.

Smearing of the d_{z^2} orbital, an s orbital, and all three p orbitals results in the five hybrids of *trigonal bipyramidal* sym-

FIGURE 4-27
The four dsp^2 square planar hybrid orbitals.

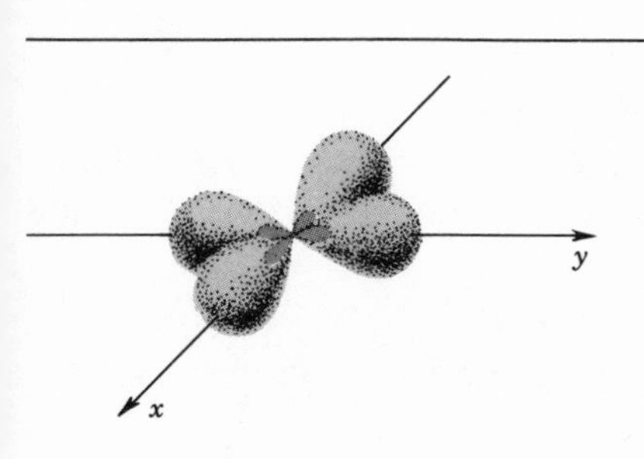

metry. These dsp^3 hybrids are illustrated in Fig. 4-29*a*. Normally the three hybrids in the xy plane are separated by 120°, while the remaining two are perpendicular to the xy plane. A maximum of five σ bonds can be constructed with these.

Six *octahedral* d^2sp^3 hybrids are formed when the $d_{x^2-y^2}$, d_{z^2}, an s, and the three p orbitals are mixed. These lie along the x, y, and z axes in plus and minus directions and can be used in construction of six σ bonds (Fig. 4-30*a*).

NOBLE-GAS AND INTERHALOGEN COMPOUNDS

4-13 With the discovery in 1962 of stable compounds of the higher noble gases, XeF_2, XeF_4, and XeF_6, chapters of many textbooks were immediately rendered out of date. For many years students had been taught that the gaseous elements He, Ne, Ar, Kr, Xe, and Rn were "chemically inert," "formed no chemical compounds whatsoever," and "remained aloof from other elements." The announcement of the synthesis of these compounds surprised chemists everywhere, for the unusual stability of the "closed-shell" noble-gas configurations and the resistance of these atoms to chemical change had long been accepted facts.

Theoreticians sharpened their pencils and shortly several different explanations of the bonding in these compounds were offered. Two of these involved no dramatic new ideas but simply pointed out the similarity between the new compounds and the well-known *interhalogens* (compounds of halogens with other halogens) for which reasonable bonding theories already existed. Figure 4-28 illustrates with line drawings the shapes of a few molecules of interest: the T-shaped interhalogen BrF_3, the linear XeF_2, the pyramidal BrF_5, the square planar XeF_4, and the octahedral XeF_6.†

One of the suggested explanations calls for hybridization of orbitals on the central atom; therefore we begin by considering the ground-state electronic configurations of bromine and xenon.

†Experimentally it is known that XeF_6 is slightly distorted from a perfect octahedral shape. Research on the configuration of this molecule is still in progress.

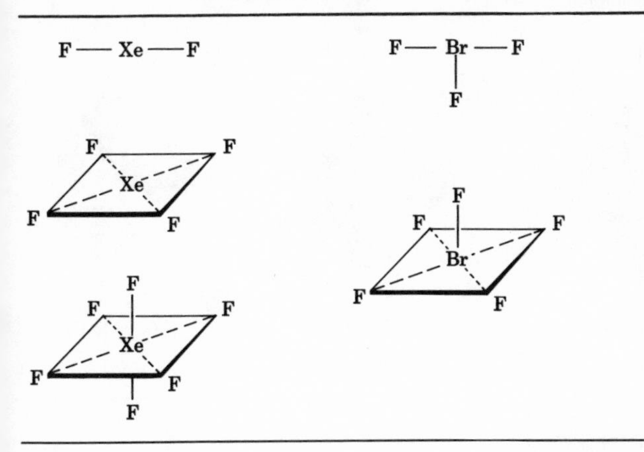

FIGURE 4-28
The geometries of the three xenon fluorides and related interhalogens.

Br ($Z = 35$): $1s^2\ 2s^2\ 2p^6\ 3s^2\ 3p^6\ 3d^{10}\ 4s^2\ 4p^5\ [5s^0\ 5p^0\ 4d^0]$
Xe ($Z = 54$): $1s^2\ 2s^2\ 2p^6\ 3s^2\ 3p^6\ 3d^{10}\ 4s^2\ 4p^6\ 4d^{10}\ 5s^2\ 5p^6$ $[6s^0\ 6p^0\ 5d^0]$

The bracketed empty orbitals are those next highest in energy in the isolated atoms.

We shall call on the lowest-energy *empty d* orbitals in Br and Xe to mix with the outermost filled *s* and *p* orbitals in forming the trigonal, bipyramidal dsp^3 hybrids illustrated in Fig. 4-29. Assignment of the seven external Br electrons as shown allows formation of three mutually perpendicular σ bonds by pairing with the lone $2p$ electron on fluorine. Two lone pairs (shaded orbitals) remain in the trigonal plane. Now, as we learned earlier in our discussions of H_2O, NH_3, and CH_4, electron-pair repulsions decrease in the order *lp-lp* > *lp-bp* > *bp-bp*. Accordingly we should expect the bond angles in BrF_3 to be less than 90°, and (comforting as it is) experiment shows that the bond angles are approximately 87°.

For XeF_2 we assign the outer eight electrons to the hybrids as shown, so that the bond formation is consistent with the observed linear structure. The three lone pairs in the trigonal

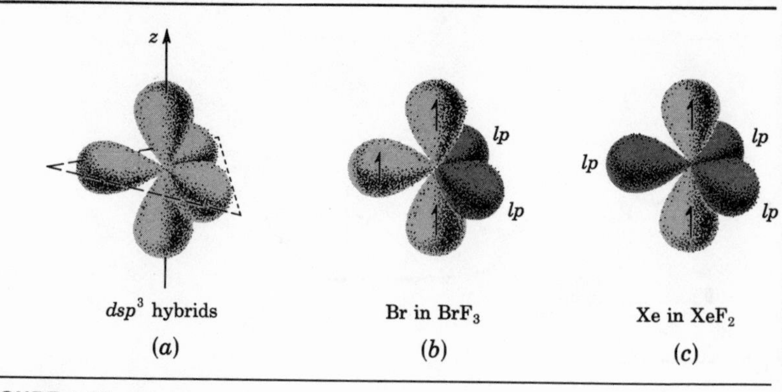

FIGURE 4-29
The hybrid model for BrF_3 and XeF_2. (*a*) The five trigonal bipyramidal *dsp^3* hybrid orbitals; (*b*) the electron configuration of Br in BrF_3; (c) the electron configuration of Xe in XeF_2.

plane exert equivalent repulsion on both bond pairs; no distortion from linearity occurs.

For BrF_5 and XeF_4 construction of d^2sp^3 octahedral hybrids is indicated. Figure 4-30 shows the allocation of electrons appropriate for formation of pyramidal BrF_5 and square planar XeF_4. Again experiment reveals the distortional effect of *lp-bp* repulsion in BrF_5; the four bonds represented as coplanar in Fig. 4-30*b* are bent slightly upward away from the lone pair. As expected, no such distortion is observed in XeF_4.

Explaining the existence of XeF_6 with a hybrid model is more difficult, for we have two too many electrons! The Xe atom in a d^2sp^3 state (Fig. 4-30*c*) has only four unpaired electrons. If we postulate that two Xe electrons, one from each lone pair, get excited to some higher empty Xe orbital (perhaps the 6*s*), then we have six unpaired electrons in the hybrids ready to form bonds with electrons on six fluorine atoms, and thus a possible structure for XeF_6. However, the excessive amount of energy needed to bring about such excitations and to promote electrons to the hybrid states shown in Figs. 4-29

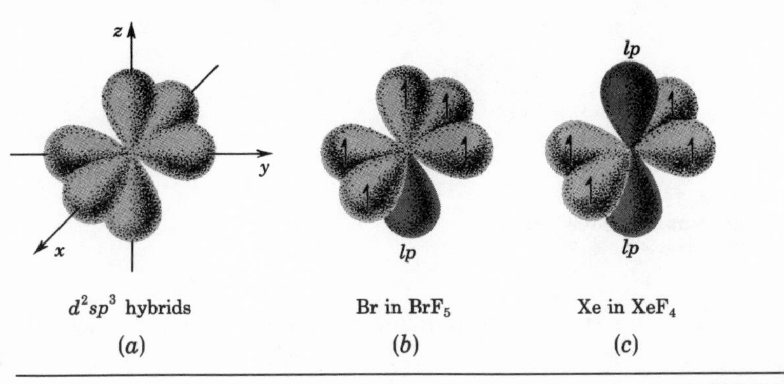

FIGURE 4-30
The hybrid model for BrF_5 and XeF_4. (*a*) The six octahedral d^2sp^3 hybrids; (*b*) the electron configuration of Br in BrF_5; (c) the electron configuration of Xe in XeF_4.

and 4-30 has been the basis for considerable criticism of the hybrid model for the interhalogens and xenon halides. Multicenter molecular orbital approaches, which we will examine next, do not require such excitations to explain the existence of these molecules and are consequently currently the favored description.

In the simplest version of these we consider the construction of a three-centered σ bond, i.e., a molecular orbital extending over three atoms. Figure 4-31 illustrates the proper orientation of orbitals for the *MO*s in linear XeF_2. Two fluorine p_z orbitals, each with one electron, overlap in a straight line with the filled $5p_z$ on Xe. Smearing of all three atomic orbitals yields three *MO*s, one that is bonding with respect to the isolated orbitals, one antibonding, and the third *nonbonding*. A nonbonding molecular orbital (*NBMO*) has an energy like those of the isolated atomic orbitals, and no bond stability is gained or lost when an electron occupies it. Consequently, if the four electrons originally in the isolated atomic orbitals are assigned two each to the *BMO* and

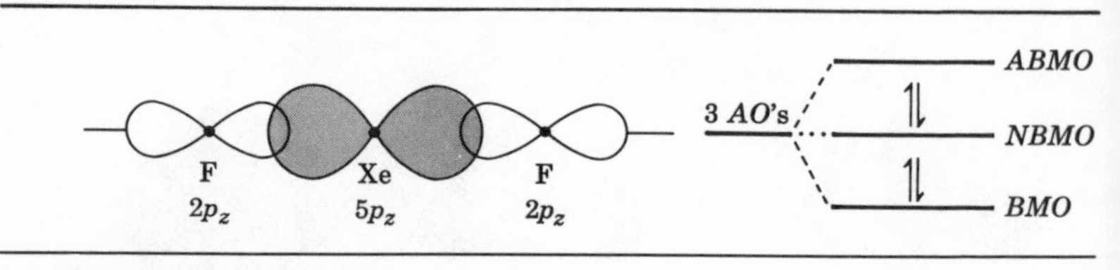

FIGURE 4-31
The three-center–four-electron *MO* model of XeF_2.

NBMO, a small net stabilization results, roughly equivalent to that of one σ bond *over all three atoms* or one-half a σ bond between Xe and each fluorine. Thus again a weak bond energy is predicted.

The structures of XeF_4 and XeF_6 (and of all the interhalogens) are explained by postulating the formation of mutually perpendicular delocalized σ *MO*s. XeF_6, for example, would have these three-center *MO*s along the x, y, and z directions.

In a more accurate treatment of XeF_6 the delocalization is considered to extend in three dimensions over the whole molecule and orbitals other than p orbitals are permitted to join in. The energy-level diagram resulting is shown in Fig. 4-32. The center lines represent calculated molecular orbital energies, and the dotted lines connect these to the atomic orbitals on Xe and F which were mixed in the *MO* formation. In this model the 5s orbital on Xe combines with 2p orbitals on all F atoms to yield nondegenerate symmetric bonding and antibonding *MO*s labeled a_{1g} and a_{1g}^* in the diagram;† the 5p orbitals on Xe combine with 2p's on F atoms to yield triply degenerate states called t_{1u} and t_{1u}^*, and the empty $5d_{z^2}$ and $5d_{x^2-y^2}$ on Xe combine with 2p's on F atoms to give doubly degenerate states e_g and e_g^*. The $5d_{xy}$, $5d_{xz}$, and $5d_{yz}$ on Xe remain nonbonding and are labeled t_{2g} in the diagram. The eight external electrons on Xe and one from each F atom are assigned to these levels in an *Aufbau* process as shown in Fig. 4-32. Of the 14 electrons occupying XeF_6 *MO*s in this diagram all but

†a_{1g}, e_g, t_{1u}, and t_{2g} are symbols derived from group theory to describe the symmetries of molecular orbitals in polyatomic molecules.

two (those in the a_{1g}^*) have gone to a state of lower energy with respect to the atomic orbitals. The antibonding pair in a_{1g}^* could at most cancel out only one of the bonding pairs in the t_{1u} orbitals. Consequently the molecule is stable.

At present both *MO* approaches and pseudohybrid theories are being tested on noble-gas compounds.

EXERCISES

1 Two possible geometries for the transient molecule N_2H_2 are illustrated below. Since its structure has not yet been deter-

FIGURE 4-32
Delocalized *MO* diagram for octahedral XeF_6.

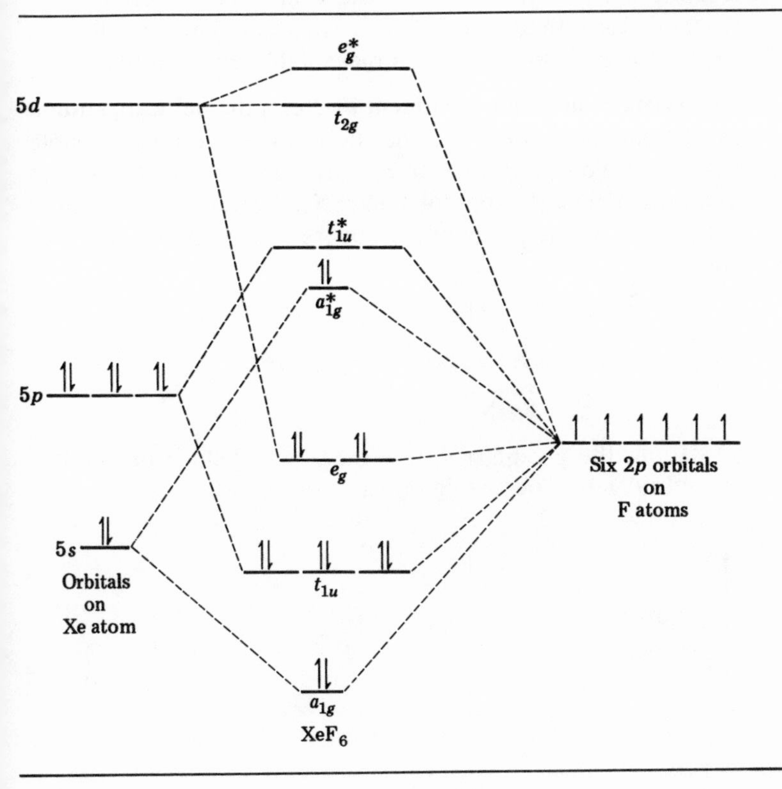

mined experimentally, we will first make some assumptions: that all atoms are in the same plane, that bond angles are all 90°, and that the molecule has no unpaired electrons.

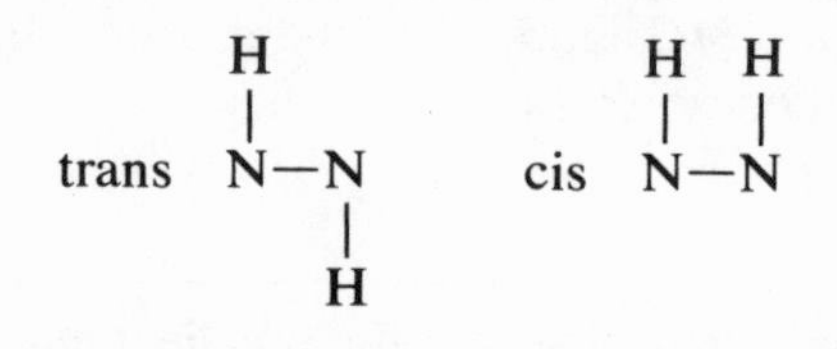

We will now try to make some predictions to assist the experimentalists.

Without invoking hybridization, propose an orbital picture of this molecule. How many σ and how many π bonds would be expected and where are their locations? Would both cis and trans forms be stable, or would the molecule freely change from one form to another? Under what conditions would you expect the molecule to have a measurable dipole moment?

2 Without invoking hybridization, propose orbital pictures for the bonding in the molecules below. Predict their probable shapes, all bond types (σ or π), any unpaired electrons, and dipole moments. (Except for square S_4, the line diagrams show only *which* atoms are bonded, not the geometry of the molecule.)

OF_2 P_2 H_2S AsH_3 H—O—O—H

S—S
| |
S—S

S—S—S—S

3 From the geometries described here deduce the probable hybridization occurring on each starred atom.

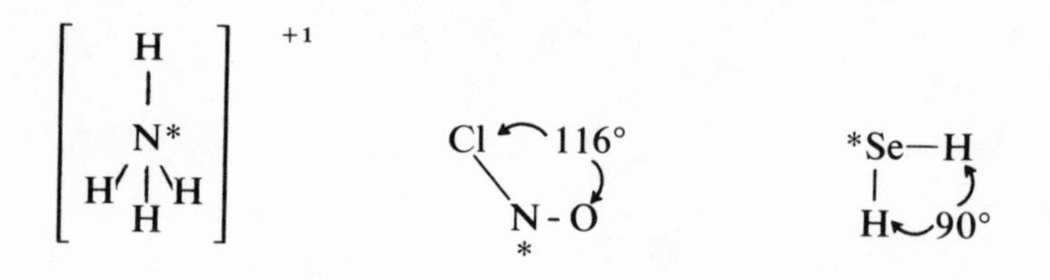

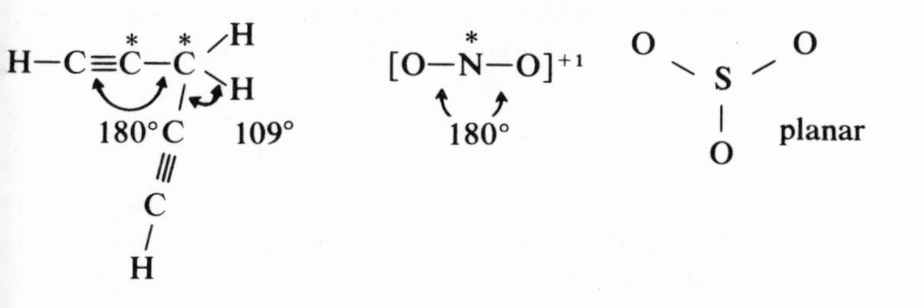

4 The CO_2 molecule is linear, with the carbon atom between the oxygens. Discuss the probable bonding in this molecule.

5 The allene molecule C_3H_4 has the line structure shown below.

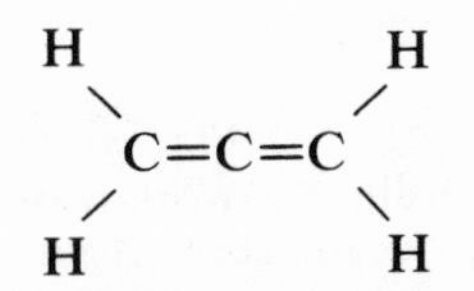

What hybridization is required for the central carbon atom? For terminal carbon atoms? Draw an orbital bonding diagram for the molecule showing why the four hydrogen atoms are not all in the same plane.

6 Which molecule would have the weakest bond? Why?

Li_2 Rb_2 H_2 K_2 Na_2

7 Which molecule would have the shortest bond distance?

Cl_2 I_2 Br_2 F_2

8 In the group NO^{2+}, NO^+, NO, NO^- which species would have the strongest bond? Which the shortest bond length?

9 Write formulas for four molecules or molecule-ions that have $BO = \frac{1}{2}$.

10 Write formulas of two neutral molecules that are iso-electronic with NO^+.

11 Which of the following possible species in the earth's ionosphere would you expect to be most stable (strongest bond) and which least stable?

$ArCl^+$ OF^+ NO^+ PS^+ SCl^+

12 Explain why the positive ion of NO_2 is linear, the neutral molecule bent, and the negative ion even more bent, as shown in the diagrams below.

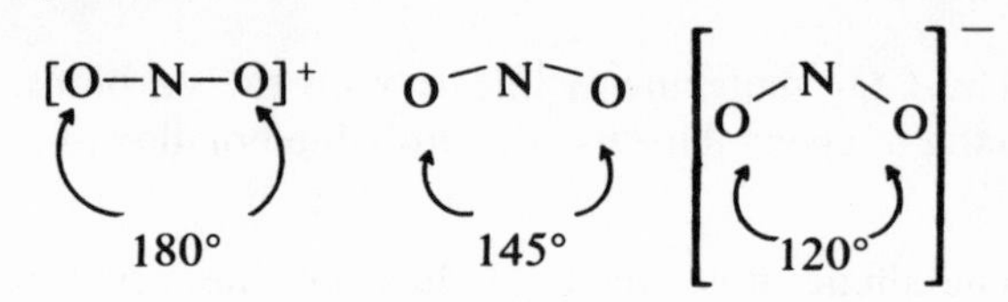

13 Suggest a possible orbital structure in terms of hybrids for the noble-gas compound $XeOF_4$.

14 Figure 4-33 shows one representation of the tetrahedral CH_4 molecule. The four H atoms are located at the four corners of a cube, and the carbon atom is at the origin of the coordinate system. Express the bond dipole-moment vectors in terms of their components along the x, y, and z axes and show that their sum is zero, i.e., that the resultant dipole moment of CH_4 is zero.

15 The nitric acid molecule has the planar framework

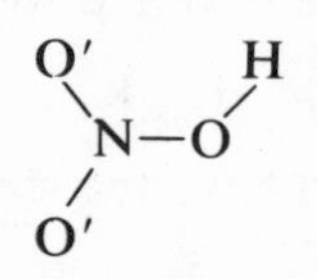

where angle O′—N—O′ is 130° and angle O′—N—O is 115°. Angle N—O—H is not well established but is known to be not 180°. The N—O distance is longer than the equivalent N—O′ distances, and there is some probable restricted rotation of the OH group with respect to the NO_2 group. Propose an orbital model for the molecule consistent with all these facts.

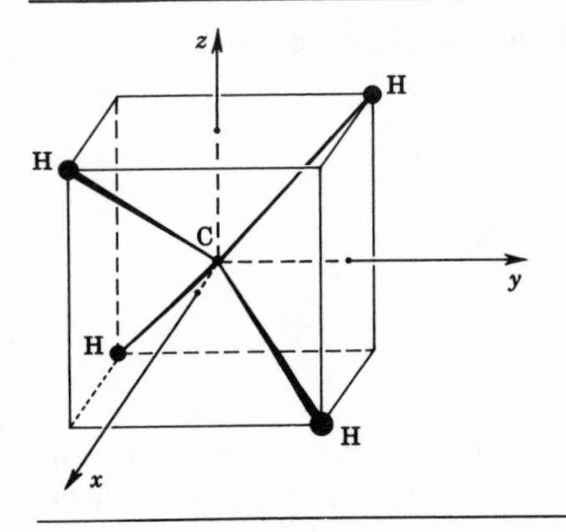

FIGURE 4-33
A representation of the tetrahedral CH_4 molecule.

16 Consider the molecule SF_4, in which the sulfur atom is bonded to four fluorine atoms. How many *lone pairs* of electrons must be located immediately around the sulfur atom? How many *bond pairs* of electrons? Refer to Figs. 4-29 and 4-30 and deduce the probable hybridization of the sulfur atom in SF_4. What are possible geometries for SF_4?

17 With what neutral fluoride molecule is the molecule-ion ClF_4^+ isoelectronic? What are its possible geometries?

18 I_3^- is an interhalogen molecule-ion of linear geometry. Using first the three-center molecular orbital approach and then one of the hybrid models, explain why this molecule-ion is stable.

19 The molecule BCl_3, as mentioned earlier, has no resultant dipole moment, even though the individual B—Cl bonds are polar. What must be the geometry of the molecule? Draw an orbital picture for the molecule consistent with your geometry showing hybridization on the boron atom and the location of all chlorine electrons. Why is BCl_3 called an "electron-deficient" molecule?

20 The molecule borazole $B_3N_3H_6$ has the framework:

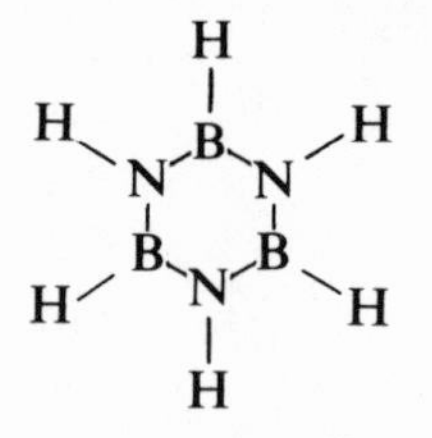

Discuss the bonding in this molecule.

21 In the diamond form of solid carbon, atoms are held in tetrahedral arrays as shown in (*a*) below. In the graphite form of solid carbon, flat layers of atoms in a chicken-wire pattern (*b*) are stacked to form the three-dimensional solid. After reviewing discussions of bonding in CH_4 and C_6H_6 given in the text, try to rationalize what types of bonds are present in diamond and in the layers of graphite.

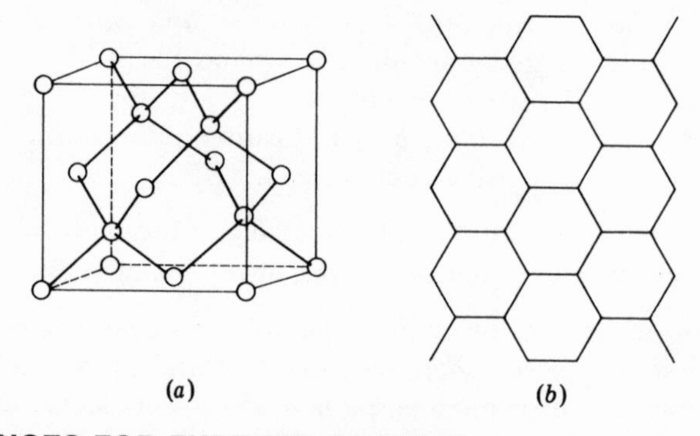

(*a*) (*b*)

REFERENCES FOR FURTHER READING

GENERAL

1 Heslop, R. B., and P. L. Robinson, "Inorganic Chemistry," 3d ed., chaps. 5–7, Elsevier Publishing Company, Amsterdam, 1967.

2 von Hippel, A. R., Molecular Designing of Materials, *Science*, **138,** 91 (1962).

3 Coulson, C. A., "Valence," 2d ed., sec. 6.3, Oxford University Press, New York, 1961.

4 Halliday, D., and R. Resnick, "Physics for Students of Science and Engineering," 2d ed., part II, sec. 30-4, John Wiley & Sons, Inc., New York, 1962.

5 Benfey, T., Geometry and Chemical Bonding, *Chemistry*, **40,** 21, May (1967).

ORBITAL SHAPES

6 Noller, C. R., A Physical Picture of Covalent Bonding and Resonance in Organic Chemistry, *J. Chem. Educ.*, **27,** 504 (1950). See also *J. Chem. Educ.*, **32,** 23 (1955).

7 Cohen, I., The Shape of the 2*p* and Related Orbitals, *J. Chem. Educ.*, **38,** 20 (1961).

8 Ogryzlo, E. A., and G. B. Porter, Contour Surfaces for Atomic and Molecular Orbitals, *J. Chem. Educ.*, **40,** 256 (1963).

9 Wahl, Arnold C., Molecular Orbital Densities: Pictorial Studies, *Science*, **151,** 961 (1966).

10 Cohen, I., and Janet Del Bene, Hybrid Orbitals in Molecular Orbital Theory, *J. Chem. Educ.*, **46,** 487 (1969).

11 Holmgren, Stephen L., and James S. Evans, Accurate Contours for sp^{α} Hybrid Orbitals, *J. Chem. Educ.*, **51,** 189 (1974).

BOND ENERGIES AND TREND PREDICTIONS

12 Howald, Reed A., Bond Energies in the Interpretation of Descriptive Chemistry, *J. Chem. Educ.*, **45,** 163 (1968).

13 Heald, Raymond R., Isoelectronic Principle, *Chemistry*, **46,** 10, June (1973).

14 Benson, Sidney, Bond Energies, *J. Chem. Educ.*, **42,** 502 (1965).

15 Bent, Henry A., Isoelectronic Systems, *J. Chem. Educ.*, **43,** 170 (1966).

LONE-PAIR ELECTRONS AND THEORIES BASED ON ELECTRON REPULSION

16 Fowles, G. W. A., Lone Pair Electrons, *J. Chem. Educ.*, **34,** 187 (1957).

17 Dahl, Peter, The Valence-Shell Electron-Pair Repulsion Theory, *Chemistry*, **46,** 17, March (1973).

18 Gillespie, R. J., The Electron Pair Repulsion Model for Molecular Geometry, *J. Chem. Educ.*, **47,** 18 (1970).

NOBLE-GAS COMPOUNDS

19 Ward, R., Would Mendeleev Have Predicted the Existence of XeF_4?, *J. Chem. Educ.*, **40,** 277 (1963).

20 Moody, G. J., A Decade of Xenon Chemistry, *J. Chem. Educ.*, **51,** 628 (1974).

21 Hyman, Herbert H., The Chemistry of the Noble Gases, *J. Chem. Educ.*, **41,** 174 (1964).

22 Kaufman, Joyce J., Bonding in Xenon Hexafluoride, *J. Chem. Educ.*, **41,** 183 (1964).

23 Chernick, C. L., Chemical Compounds of the Noble Gases, *Rec. Chem. Prog.*, **24,** 139 (1963).

BONDING IN IONIC, METALLIC, AND MOLECULAR SOLIDS

5

IONIC BONDING; STABILIZATION OF IONS IN CRYSTALS

5-1

Earlier we said that atoms whose external electronic configurations were close to those of the noble gases (atoms on the far right or far left of the periodic table) tend to attain the rare-gas configuration by losing or gaining electrons, as appropriate. However, if we examine the energy relationships accompanying such processes for the isolated atoms, this statement at first seems unreasonable. Consider a system containing 1 mol of gaseous Na atoms separated from 1 mol of gaseous Cl atoms. If we *supply* to the system an amount of energy I_1, the first ionization energy of Na, we can effect the process:

$$I_1 + \mathrm{Na} \rightarrow \mathrm{Na}^+ + e^-$$

For 1 mol of Na atoms I_1 is 496 kJ.

If 1 mol of electrons is supplied to the Cl atoms in the system, energy equal to the electron affinity (EA) of Cl is released. The electron affinity of Cl is 348 kJ/mol.

$$e^- + \mathrm{Cl} \rightarrow \mathrm{Cl}^- + EA$$

Thus to create ions in the gas phase, with no interaction between ions as yet, we must do 148 kJ of work:

$$148 \text{ kJ} + \mathrm{Na} + \mathrm{Cl} \rightarrow \mathrm{Na}^+ + \mathrm{Cl}^-$$

Therefore in Fig. 5-1 the state of the system of isolated ions is shown as a less stable state compared to that of the isolated

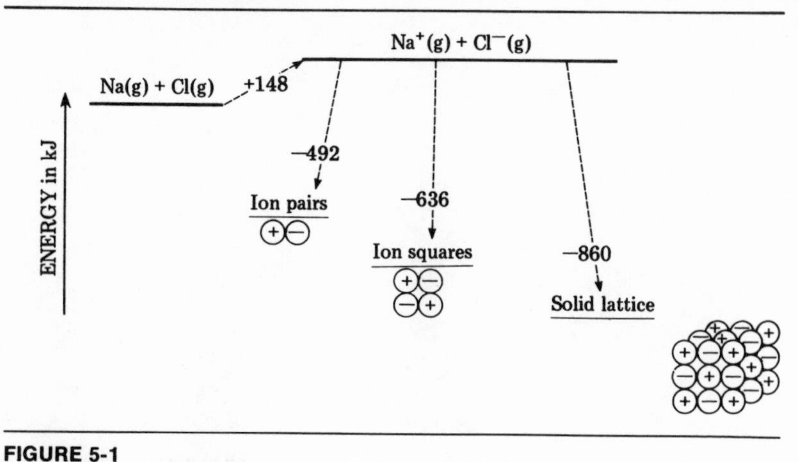

FIGURE 5-1
Stabilization energies for a system of 1 mol of Na^+ ions and 1 mol of Cl^- ions brought together as ion pairs, ion squares, and as a solid. Only a small portion of the solid lattice is shown.

atoms. What rescues us from this unpleasant situation is the fact that ions, as charged particles, attract one another according to Coulomb's law, gaining *per single ion pair* a classical electrostatic stabilization energy given by

$$E_{ip} = \frac{(+z_c)(-z_a)e^2}{r}$$

where z_c and z_a are positive integers characteristic of the positive ion (the cation) and the negative ion (the anion), $+z_c e$ is the total charge on the cation expressed as a multiple of the electron charge, and r is the distance between the ion centers. As the separation between ions decreases, this stabilization energy obviously increases in magnitude. Now although we treat the ions as point charges when discussing their attraction, we must also recognize their finite size. There is a distance of closest approach d (Fig. 5-2), determined when the outer filled orbitals of the ions begin bumping one another and repelling.

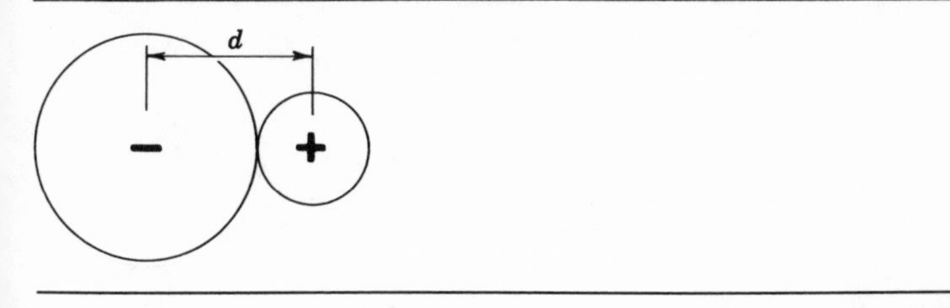

FIGURE 5-2
Ion-pair formation.

The peculiar stability of the noble-gas configurations allows us to consider the ions as hard spheres, at least to a first approximation. Balance between electrostatic attraction and repulsion by the electrons causes an equilibrium interionic distance to be established, like the equilibrium distance between atoms in molecules. For an Na^+-Cl^- pair the gas phase internuclear distance d is 0.2814 nm. With the conversion factor $e^2/1\text{ nm} = 138.5\text{ kJ}/N$, we can express the single ion-pair energy as

$$E_{ip}(\text{NaCl}) = \frac{(+1)(-1)e^2}{0.2814\text{ nm}} = \frac{-492.1\text{ kJ}}{N}$$

where N is Avogadro's number.

Our 1 mol of Na^+ and 1 mol of Cl^- particles are sufficient to form 1 mol or N ion pairs. Thus the total stabilization energy due to ion-pair formation, U_{ip}, can be calculated as

$$U_{ip} = NE_{ip} = -492.1\text{ kJ}$$

enough to bring our system into a state *more* stable than the isolated atoms (Fig. 5-1).

If two ion pairs cluster together to form an ion square (Fig. 5-3), the stabilization energy increases. For a single ion square we have

$$E_{is} = \frac{4(+1)(-1)e^2}{d} + \frac{(+1)(+1)e^2}{\sqrt{2}d} + \frac{(-1)(-1)e^2}{\sqrt{2}d}$$

$$= \frac{2(1.293)e^2}{d} = 2(1.293)E_{ip}$$

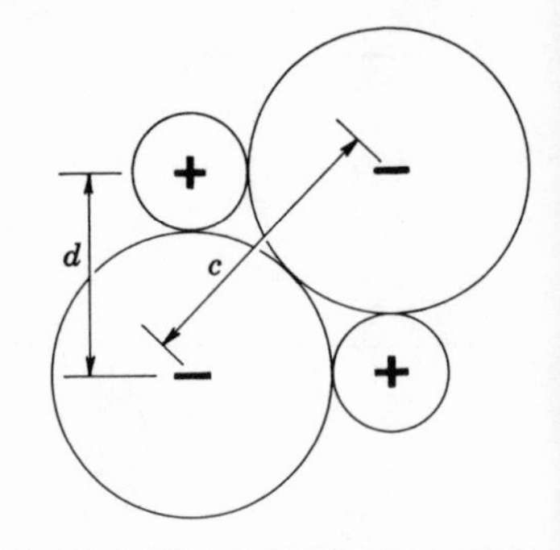

FIGURE 5-3
Ion-square formation. $c = \sqrt{2}d$.

In our assembly of ions we have $N/2$ such ion squares possible, so that the total stabilization energy due to ion-square formation, U_{is}, is

$$U_{is} = \frac{N}{2} E_{is} = 1.293 N E_{ip} = -636 \text{ kJ}$$

When *all* the ions in our assembly are brought together in a single cluster, conceivably built by stacking ion squares in all directions, the result is the macroscopic crystalline solid NaCl, a three-dimensional *lattice* of ions. As one might expect, the resulting stabilization is even greater than that in the ion-square case. The stabilizing energy U_L is known as the lattice energy of the solid and is always expressible in terms of a constant A_M (known as the Madelung constant), the Avogadro number N, and the single-ion-pair energy:

$$U_L = A_M N E_{ip}$$

The Madelung constant for a solid is analogous to the constant 1.293 which we obtained for ion squares and is computed in a similar manner, by summing interactions in all directions in the solid until the sum converges. The Madelung constant will thus depend upon *how* the ions cluster together, that is, upon the geometry of the crystal. For solids which crystallize like NaCl (square upon square to form a cubic structure) the

Madelung constant is 1.7476. Thus U_L for our mole of NaCl is −860 kJ. Figure 5-1 shows the total stabilization gained.

For a more complete picture of the energy effects accompanying the formation of an ionic solid from the elements *in their natural states* we would indicate in Fig. 5-1 the energy necessary to convert the solid metal to gaseous atoms (the heat of sublimation of the metal) and the energy required to create the anion-to-be (in the case of chlorine, the dissociation energy of molecular Cl_2 gas).

The lattice energy represents to a first approximation the binding energy of the ions in the crystal. More rigorous treatments account for the fact that ions are not really point charges or even hard spheres, but are slightly squashable species; these effects, however, introduce relatively small corrections to the electrostatic energy.

The magnitude of the lattice energy has significant influence on certain physical properties of a solid, such as its melting point and its solubility. NaCl and KCl, which have like crystal structures, have melting points 801 and 768°C, respectively, the small decrease being due to increase in d and consequent decrease in U_L. MgO, also of cubic structure, melts at 2800°C, the difference due partly to decrease in d but primarily to increase in ionic charge. From examination of the single-ion-pair formula we see that E_{ip} (and consequently U_L) increases fourfold when ionic charge increases from 1 to 2.

Solubility differences among salts may be related to the differences between their lattice energies and the so-called hydration energies of ions in solution. Positive ions in a water environment may gain a stabilization energy by attracting to them the lone-pair oxygen electrons of water molecules. Indeed this electrostatic binding is so strong that, as the cations wander through the solution, water molecules are dragged along with them. Anions achieve a similar stabilization by attraction to the hydrogen atoms in the water molecule. Because of unequal sharing of electrons in the O—H bond (O is more electronegative), the H atom is almost a bare proton, a bare positive charge sitting on the end of the OH sigma *MO*. The anion moves up to it and achieves stabilization through electrostatic attraction.

At the surface of a crystal immersed in water there rages a contest between lattice stabilization and hydration stabilization. If the latter is greater, the crystal dissolves.

The possibility of hydration stabilization accounts for the greater water solubility of most ionic compounds compared to many covalent molecules.

SIZE OF IONS

5-2 Thus far we have implied that ions, regarded as hard spheres in crystalline solids, have a measurable size; yet this is not strictly true. We *can* measure the distance between ion centers in a solid by means of x-ray diffraction (see Chap. 1). With a known wavelength of x-radiation, measurement of the angles through which the beam is bent allows us to calculate the distance between the ion centers. To get a measure of the radius of the hard sphere, we must resort to more devious means. Consider the solid LiBr, and assume that we have learned from x-ray studies how the ions are positioned in the crystal (in our familiar cubic structure) and the distances between ion centers. The Li^+ ion, with only two electrons, would be smaller than Br^- with 36 electrons, so that we could safely assume that the packing in LiBr was largely determined by the Br^- ions, as shown in Fig. 5-4. One-half the measurable distance d can be taken as the radius of the bromide ion. One may

FIGURE 5-4
Packing of spheres in LiBr and KBr.

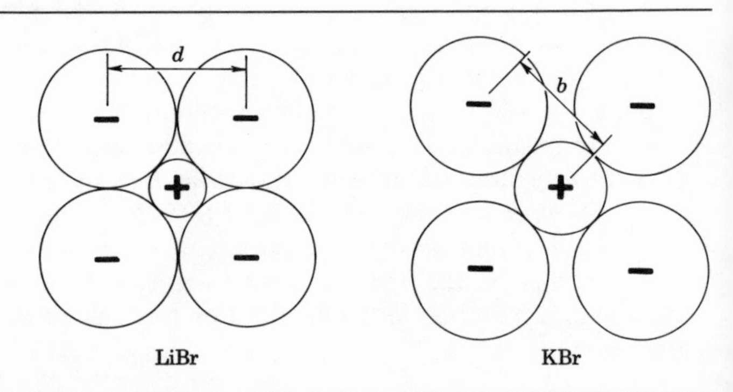

TABLE 5-1
Some Representative Averaged Ionic Radii in Nanometers

								H^-	0.208
Li^+	0.068	Be^{2+}	0.035			O^{2-}	0.140	F^-	0.136
Na^+	0.097	Mg^{2+}	0.067	Al^{3+}	0.051	S^{2-}	0.184	Cl^-	0.181
K^+	0.133	Ca^{2+}	0.099			Se^{2-}	0.198	Br^-	0.195
Rb^+	0.147	Sr^{2+}	0.112			Te^{2-}	0.221	I^-	0.216
Cs^+	0.167	Ba^{2+}	0.134						

Ti^{2+}	V^{2+}	Cr^{2+}	Mn^{2+}	Fe^{2+}	Co^{2+}	Ni^{2+}	Cu^{2+}	Zn^{2+}
0.090	0.088	0.084	0.080	0.074	0.073	0.069	0.072	0.074

Bi^{3+}	0.096	Sc^{3+}	0.081	Mn^{3+}	0.066	Si^{4+}	0.042
Cr^{6+}	0.052	Cu^+	0.096	Ce^{4+}	0.094	Ti^{4+}	0.068

then, through a similar study of KBr, where the ions are of comparable size (Fig. 5-4), determine r_{K^+} by subtraction of r_{Br^-} from the measurable distance b. The amazing thing is that these radii, so determined, appear to be reasonably constant from compound to compound. For example, if one determined r_{I^-} with solid LiI and proceeded through KI to get r_{K^+}, the value obtained is very close to that of the bromide study. Through paths like these taken over most of the periodic table it is possible to set up a self-consistent set of *averaged* ionic radii, a representative sample of which is shown in Table 5-1. The constancy of ionic radii enables us to make many valuable predictions about crystal structure.

TRENDS IN IONIC RADII

5-3 The major patterns apparent in Table 5-1 are easily explained in terms of orbital theory and may be classified in terms of four general rules.

RULE 1 For members of an isoelectronic series (a series of ions with the same number of electrons) the ionic radius decreases with increase in atomic number. This is simply the effect of increased nuclear attraction for the electron cloud.

$$r_{Mg^{2+}} < r_{Na^+} < r_{F^-} < r_{O^{2-}}$$

RULE 2 Ions in a family in the periodic table increase in radius as we travel down the column, a consequence of adding electrons with their most probable distance farther from the nucleus.

$$r_{F^-} < r_{Cl^-} < r_{Br^-} < r_{I^-}$$

RULE 3 For cations of the same element, the radius decreases with increasing ionic charge because there are fewer electrons to repel one another and expand the cloud charge. For example, the radius of Fe^{2+} is 0.076 nm while the radius of Fe^{3+} is 0.064 nm. The reverse holds true for anions; i.e., O^{2-} is larger than O^-.

RULE 4 Transition-metal ions of equal ionic charge (Ti^{2+} to Ni^{2+}, for example) exhibit less dramatic changes in radii with increase in atomic number (see Table 5-1). Addition of a proton to the nucleus and an electron to the *d* orbitals seems to be an almost compensatory process. (The slight variations observed, however, can be explained. See Exercise 5 in Chap. 6.)

RADIUS RATIO AND PACKING OF IONS IN CRYSTALS

5-4 The concept of radius ratio enables us to apply our tables of ionic radii to qualitative prediction of how ions are most likely to be clustered in crystals and what are most probably the formulas of certain clusters of ions, i.e., complex ions, found in solution. First we define the coordination number (CN) of a central ion as the number of ions of opposite charge actually touching it. If ion A is surrounded by three B ions in a planar trigonal configuration, its CN is 3. If its neighbors are in a square planar or tetrahedral arrangement, its CN is 4. Other arrangements are mentioned in Table 5-2.

The radius ratio R_r for a binary ionic solid CA is defined as r_c/r_a, where r_c and r_a are ionic radii of cation and anion. Now, obviously, within the hard-sphere approximation, as the cation becomes larger with respect to the anion (as R_r increases), more anions can fit around the cation; i.e., the CN of the cation may increase. Table 5-2 shows the possible CN for various radius ratios as computed from simple geometric considerations

TABLE 5-2
Radius Ratio and Coordination Number

R_r	*CN*	SYMMETRY
< 0.155	1 or 2	Linear or bent molecules
0.155–0.225	3	Trigonal planar environment of *A*
0.225–0.414	4	Tetrahedral environment of *A*
0.414–0.732	4	Square planar environment of *A*
0.414–0.732	6	Octahedral environment of *A* (example, NaCl)
0.732–1.0	8	Body-centered-cubic, or twisted-cubic environment of *A*
1.0	12	Close-packed structure of metals where all atoms are alike

about sphere packing. Armed with these, we can make simple (but not infallible) predictions. Suppose that we know that the beryllium ion in solution forms a complex ion with the fluoride ion of formula $[BeF_n]^{(n-2)-}$. Computing R_r for Be^{2+} and F^-, we obtain 0.232, which indicates a tetrahedral BeF_4^{2-} ion, perhaps fortuitously in agreement with experiment. For solid NaCl, R_r is 0.525, consistent with $CN = 6$. For CsCl we obtain an R_r of 0.934, in agreement with the *CN* of 8 known experimentally. The radius-ratio concept is generally more successful when applied to solids than to complex ions in solution, but in all cases we are limited by the fact that ionic radii themselves vary somewhat with coordination number.

IONIC POTENTIAL AND PARTIAL COVALENCY

5-5 A second useful (but again not infallible) concept involving ionic radii is that of ionic potential, which enables us to rationalize discrepancies between theoretical predictions based on perfect ionic character and experimental results.

The ionic potential† (*IP*) of an ion is defined as

$$IP = \frac{z_c}{r_c}$$

†The term ionic potential should be distinguished from the misused term *ionization potential*. Unfortunately, *ionization potential* is the name given to the *ionization energy* in many texts. Ionization energy (*I*) is the energy (in kilojoules per mole) necessary to remove an electron from an atom or molecule; ionic potential (*IP*) is a defined, nonexperimental quantity with the true dimensions of an electrostatic potential, i.e., charge per unit distance.

where z_c is the ion charge expressed as an integral multiple of e, the electron charge.

Cations with a large ionic potential have considerable *polarizing power*, or ability to distort neighboring electron clouds toward them, thus inducing *partial covalence*. Consequently, comparison of ionic potentials allows us to make qualitative comparisons of degrees of covalence. For example, the oxide of sodium, Na_2O, is considered to be more ionic than its neighbor MgO. The *IP* for Na^+ is 1/0.95 or 1.05, while that of Mg^{2+} is 2/0.65 or 3.08.

METALS

5-6 A survey of the observed properties of metals indicates that the nature of the forces holding together the atoms in a crystal must indeed be unusual. Metals are ductile and malleable and have high tensile strength; they are excellent conductors of both heat and electricity; they are opaque to light but lustrous; they can be mixed with other metals or nonmetals to yield alloys with variable properties. All of these characteristics are direct consequences of the peculiar bonding/crystal structure combination that metals possess.

Most metals crystallize in one of the packing structures shown in Fig. 5-5. In the *body-centered-cubic* (bcc) structure (Fig. 5-5*a*) we see that only atoms along the body diagonal are in contact and that each atom has eight nearest neighbors. The density of a bcc crystal is 0.162 m/R^3, where m and R are the atom's mass and radius.

For *hexagonal close packing* (hcp) we start with a layer of atoms in contact (Fig. 5-5*b*) and cover it with a second layer of identical atoms (Fig. 5-5*c*). Note that only half of the holes in the first layer can be covered. In hcp structures, atoms in the third layer occupy sites directly over those of the first layer, and atoms of the fourth layer repeat the pattern of the second layer. Successive layers repeat layers 1 and 2. Consequently the holes indicated in Fig. 5-5*c* are never covered, and an hcp structure has a characteristic series of holes penetrating the lattice. (As a memory device: *hcp* structures have *holes!*)

In *cubic-close-packed* (ccp), also called *face-centered-cubic*

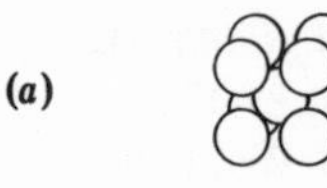

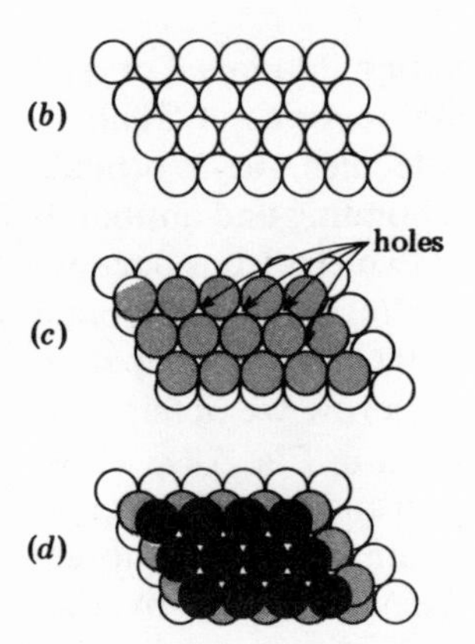

FIGURE 5-5
Typical packing of atoms in metal crystals. (*a*) Body-centered-cubic; (*b*) a single layer of close-packed atoms; (*c*) two of the repeating layers characteristic of hexagonal close-packed structures; (*d*) the three layers characteristic of cubic close packing.

(fcc), structures, layers 1 and 2 are like those of hcp structures. Atoms in the third layer, however, do not occupy sites repeating layer 1, but instead, as shown in Fig. 5-5*d*, cover the holes seen in Fig. 5-5*c*. Successive layers repeat the patterns of layers 1, 2, and 3, so that in ccp structures no holes can be seen penetrating the lattice.

Both hcp and ccp have densities of $0.176\ m/R^3$, and there are 12 nearest neighbors to every atom.

With 8 or 12 nearest neighbors and, as is the case for many common metals (Na, Ag, Au), only one electron available for bonding, localized electron-pair–bond theories do not provide a satisfactory explanation for the binding of so many atoms with so few electrons. The strength of the metallic bond precludes any explanation based on weak van der Waals forces (see Sec. 5-8).

To understand the nature of the bonding in a metal crystal we must reconsider the molecular-orbital theories of Chap. 4. In the *MO* treatment of diatomic Li_2 molecules, the 2*s* orbitals on the two atoms were merged to form bonding and antibonding *MO*s as shown in Fig. 5-6*a*, with the two electrons occupying the lower level. For three Li atoms in a line, we obtain bonding, antibonding, and nonbonding *MO*s with occupation shown in Fig. 5-6*b*. For four Li atoms in a row we obtain four *MO*s with energies and occupation shown in Fig. 5-6*c*. If we now take a giant step in our rationalization and consider a crystal of *N* (say Avogadro's number) lithium atoms in a three-dimension array, we may deduce that the *N* 2*s* orbitals overlap

FIGURE 5-6
Changes in molecular orbital energy patterns as 2, 3, 4, and *N* lithium atoms combine.

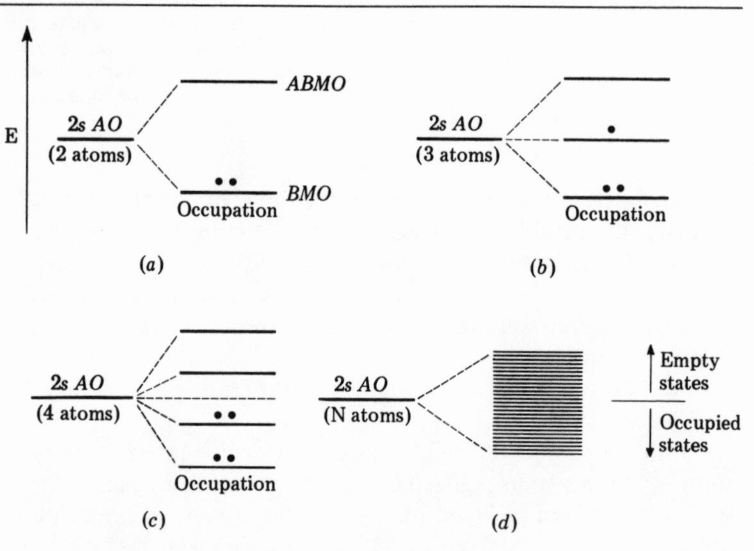

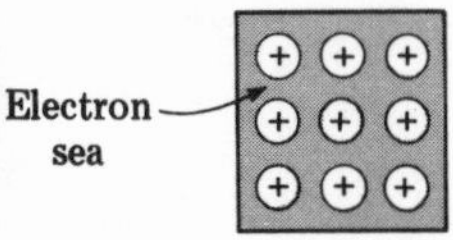

⊕ Li^+ ($1s^2$) core

Electron sea

FIGURE 5-7
Electron "sea" model of lithium metal.

to form crystal-wide *MO*s (N in number) with energies so close that an "almost" continuum (*quasi-continuum*) results (Fig. 5-6*d*). This group of closely spaced energy states is called a *band*, and the N electrons originally located on individual Li atoms now occupy the lower $N/2$ states.

The electron density diagram corresponding to this situation (Fig. 5-7) may be derived by superimposing all bonding *MO*s in the crystal, the result being a pattern of delocalization of all $2s$ electrons over the crystal around $1s^2$ cores. In more colorful language, the cores are immersed in a "sea" of electrons, or the cores are held together by an electron "glue" spread about them. These pictures of a delocalized, nondirectional bond will be useful later in explaining many metallic properties.

The strength of metallic bonds (or the effectiveness of the electron glue) is reflected in the *heats of sublimation* of metals, i.e., the enthalpy change for the endothermic process:

$$\text{M (solid)} \rightarrow \text{M (gaseous isolated atoms)} \qquad \Delta H = \Delta H_{\text{sub}}$$

Table 5-3 contains sublimation energies for alkali metals. Note

TABLE 5-3
Heats of Sublimation of Alkali Metals

METAL	ΔH_{sub} (kJ/mol)
Li	159
Na	109
K	88
Rb	84
Cs	79

that the decrease down the periodic table parallels the behavior of dissociation energies of diatomic molecules. The decrease can be attributed to the increasing size of atoms and interatomic distance, increasing size and diffuseness of contributing atomic orbitals, and consequent thinness of the electron glue.

Figure 5-8 shows how ΔH_{sub} varies across the first long row of metals. Explanation of this trend is more complex than that for alkali metals because several factors are involved: differences in crystal structure and the size of metal atoms, and the number and kind of participating electrons and their spin states. Some useful observations are (1) that bond strength increases from K to Ca (one additional *s* electron helps) and (2) that *d* electrons in transition metals do contribute something to bonding. The dip in the curve at Mn (d^5) is possibly due to the extra spin stability of the half-filled *d* shell, with electrons preferring to remain in localized orbitals rather than to participate in metal bond formation.

In metals like lithium, other upper empty orbitals in addition to the 2*s* may overlap to form empty bands. The extent of the overlap and the width of the bands depend on orbital size and interatomic distance in the crystal. Figure 5-9*a* (the analog of a diatomic potential-energy curve) shows how band width may

FIGURE 5-8
Sublimation energies of transition metals.

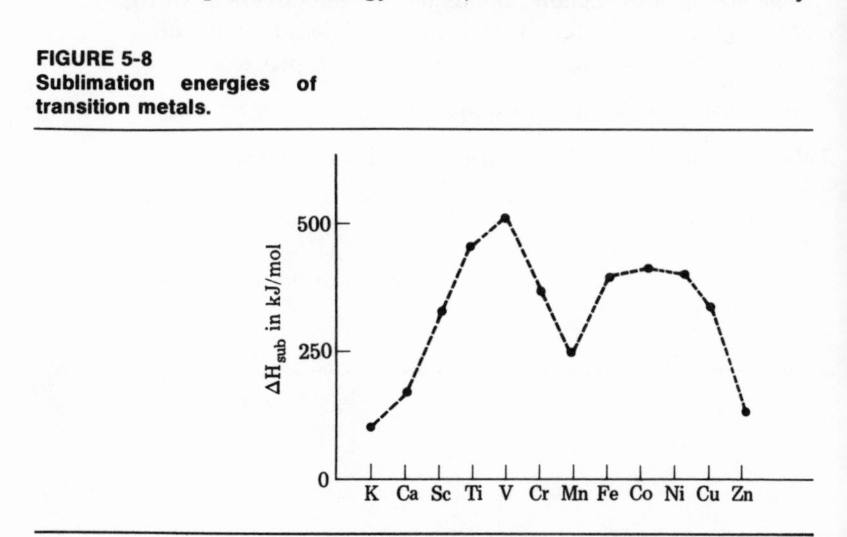

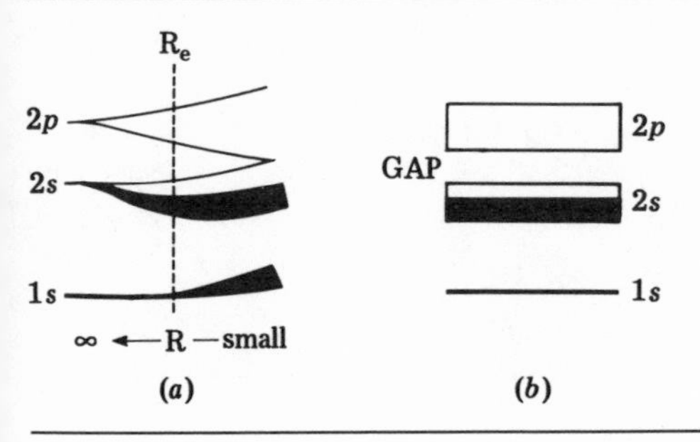

FIGURE 5-9
Band energy schemes in lithium metal. (*a*) Filled (shaded) and empty band width variation with interatomic distance; (*b*) schematic diagram of bands at crystal equilibrium interatomic distance, R_e.

vary with change in internuclear distance. For some metals the 2*s* and 2*p* bands may overlap at the equilibrium distance in the crystal. The schematic diagram (Fig. 5-9*b*) abbreviates the more complete one, showing band widths and spacings at the equilibrium distance in the Li crystal. Notice that energy *bands* replace the *discrete* energy states found in isolated atoms and that the spaces between bands represent energies *forbidden to electrons in metals*, just as energies between 1*s* and 2*s*, 2*s* and 2*p* states are forbidden to the electron in the hydrogen atom. These forbidden energies in crystals are referred to as *energy gaps*.

Now that we are armed with this much theory about metals, we will try to explain a few of the peculiar and wonderful *properties* of metals. We will first examine the mechanical properties of *ductility* and *malleability*.

Ductile metals may be easily bent or stretched into a wire without breaking; malleable metals may be pounded into a new shape. Both properties are related to the strength of the electron glue and to the ease with which crystal planes may slide over one another. A so-called *slip plane* is one along which a relatively low energy barrier exists toward displacement, so

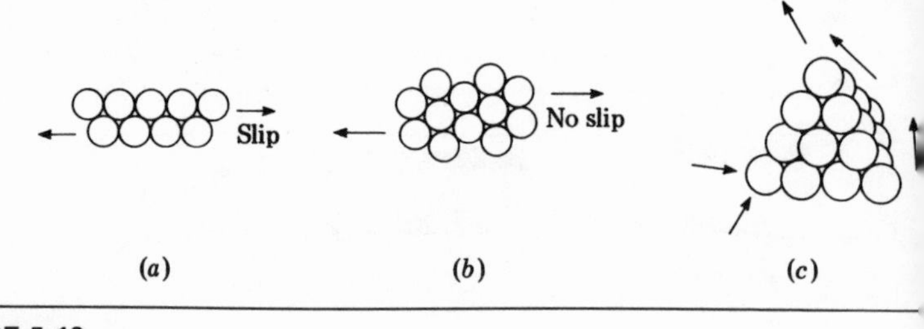

FIGURE 5-10
Slip planes (*a*) and nonslip planes (*b*). (*c*) Slip planes in ccp structures.

that under a stress one layer of atoms may easily move with respect to another to a new equilibrium position (Fig. 5-10*a* and *b*). Nondirectional bonds are necessary for this to occur without fracture. From Fig. 5-10*c* we see that ccp structures have slip planes in many directions; metals with this structure (Cu, Ag, Au, Pb) are easily deformed. The hcp structure is associated with fewer slip planes, and metals with this structure (Zn, Co, Ti) tend to be more brittle.

The high electrical conductivity of a metal can easily be explained in terms of the delocalized electron-band model. Consider the half-filled 2*s* band of lithium (Fig. 5-11*a*). The highest energy state corresponding to an exactly half-filled band is called the *Fermi level*, E_F. Above the Fermi level there are unoccupied states immediately available to electrons in the band.

Electrons in the 2*s* band are moving in all directions with equal probability. Thus there is no *net* flow in any preferred direction and hence no current. When an electric field is applied, electrons near the Fermi level under the force of the field increase their kinetic energy in the direction of the field, and a current flows. The increase in energy is possible *because* there are empty levels corresponding to higher energy *adjacent* to E_F, to which the electrons can go. Only electrons near the Fermi level are able to increase their energy and participate in conduction. The electrons deep down in the band are surrounded by fully occupied quantum states, and the Pauli exclusion principle prohibits their excitation.

To further emphasize the need for empty levels near E_F, consider beryllium metal (Fig. 5-11*b*), which *does* conduct electricity well, despite its fully occupied 2*s* band. (The configuration of the Be atom is $1s^2\ 2s^2$.) If the energy scheme appropriate for beryllium were that of Li there would be no nearby levels to which electrons could go under the influence of an electric field; hence solid Be would be an *insulator*. That the solid *does* conduct means that the 2*s* and 2*p* bands must overlap and that empty states exist adjacent to the Fermi level.

The electrical conductivity does *not* depend on the number of electrons engaged in bonding. Experiments indicate that about 1 electron per atom is effective in carrying current. The conductivity *does*, however, depend on the packing of the atoms in the metal crystal. If the current-carrying electron is a wave, it can travel through a perfectly ordered crystal lattice without resistance, like x-rays through a crystal or light waves through a grating. In a real-life metal crystal there are many *defects* (impurities, grain boundaries, vacancies) that scatter the electron waves, much as floating debris scatters waves on a lake, and consequently the conductivity is less than its value in an "ideal" metal crystal.

Electrical conductivity of a metal decreases with an increase in temperature. Again, within a wavelike model for the

FIGURE 5-11
Schematic band schemes for lithium and beryllium

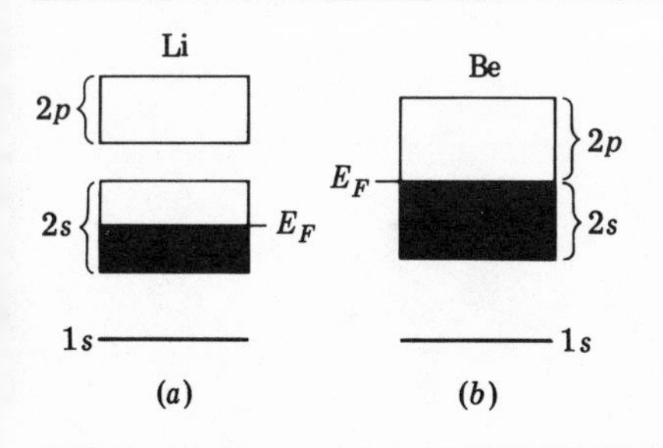

carrier electron, a suggested picture is that the electron wave collides with vibrating atoms. These lattice vibrations increase with temperature and so increasingly disturb and scatter the carrier electron wave.

Because of the nondirectional character of metallic bonding many metal atoms are not particular as to the identity of their neighbors and are willing to accept as part of the lattice any metallic atom of similar size. This accounts for the feasibility of *welding* or *soldering* processes, in which two different types of metals are fused together, and for the existence of a large class of intermetallic materials of variable composition called *alloys*.

Design of alloys and their properties may best be understood by examining how foreign metal or nonmetal atoms can be accommodated in a metal lattice, illustrated in Fig. 5-12. Foreign atoms in a *substitutional* alloy are generally other metal atoms of comparable size, while those in an *interstitial* alloy are smaller nonmetallic atoms such as C, B, or H. Note that in both cases illustrated the foreign atoms occupy randomly selected sites and thus create disorder in the basic lattice. This disorder causes many of the physical properties of alloys to differ from metals. The electrical conductivity of an alloy, for example, is lower than that of a pure metal because the electron wave carriers are scattered by the disordered foreigners in the lattice. Softer metals may be made harder by alloying. The intruder atoms, because of their different size and random locations, interfere with the slipping of planes, acting as physical barriers to slip (Fig. 5-13).

FIGURE 5-12
Basic lattice (*a*), substitutional alloy (*b*), and interstitial alloy (*c*).

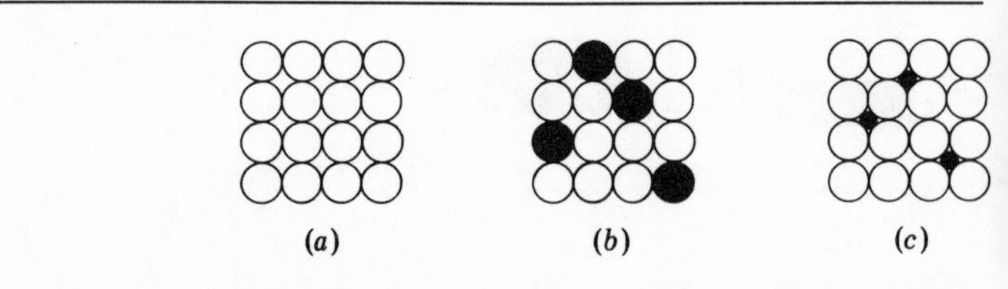

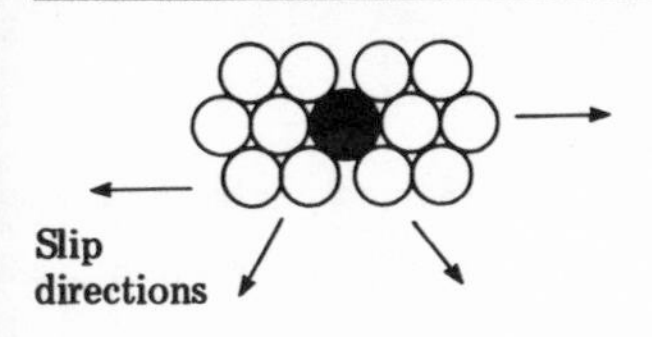

FIGURE 5-13
Effect of a foreign atom on the slip planes.

Most of the previous discussion has been based on the assumption that we have "ideal" metal crystals, free of defects and impurities. In the real world, however, metals do not solidify from a melt as one large perfect crystal but generally have one or more defects: *grain boundaries* between small crystallites, *vacancies* (where atoms missed their turn in the crystallization process), unplanned *impurities*, and *dislocations* (where whole planes get disjointed). For more details on these, see Ref. 4.

SEMICONDUCTORS

5-7 A semiconductor is one of a large class of covalent solids with a wide range of electrical conductivity values intermediate between insulators and true metals. We will demonstrate the properties of a representative semiconductor by examining the electronic structure of solid silicon, which crystallizes with tetrahedral coordination of all silicon atoms to one another. As a first approximation to a bond description assume that each silicon atom ($3s^2\,3p^2$ ground state) forms tetrahedral sp^3 hybrids containing one electron each. These overlap with similar hybrids on neighboring silicon atoms to form localized sigma-bonding *MO*s with electrons paired to make a covalent *network*, shown in two dimensions in Fig. 5-14. To each bonding *MO* there corresponds an antibonding *MO*, usually not discussed because it is under normal circumstances higher in energy, empty of electrons, and unimportant in descriptions of the ground state. Similar anti-bonding states exist in all molecules. In crystalline silicon, since the component atomic orbitals ($n = 3$) are large and diffuse, the "localized" *MO*s overlap,

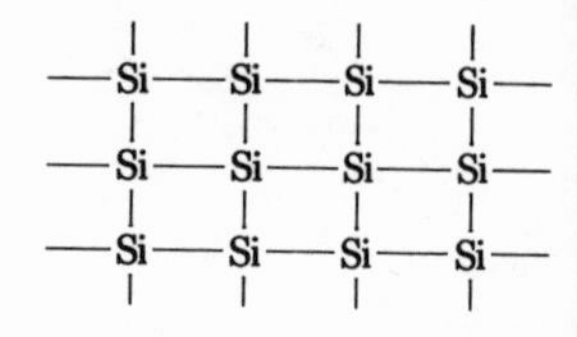

FIGURE 5-14
Two-dimensional representation of the tetrahedrally arranged silicon atoms in a silicon lattice. All bonds are covalent.

interact with one another, and, in a manner similar to that described for metals, form energy *bands* in the crystal, as illustrated in Fig. 5-15. At 0 K the lower band, called the *valence band* (corresponding to many bonding *MO* states), is completely filled, while the upper band, called the *conduction band* (composed of many antibonding states), is completely empty of electrons. Separating these bonds is a forbidden zone called the *energy gap*, E_g, where no allowed electronic states exist.

In Fig. 5-16 we contrast the band pictures of a metal, an insulator, and an *intrinsic* semiconductor such as silicon. The

FIGURE 5-15
Energy levels resulting from the overlap of (*a*) isolated *sp*³ hybrids and (*b*) many *sp*³ hybrids in close proximity to one another. (*c*) The schematic energy diagram for case (*b*).

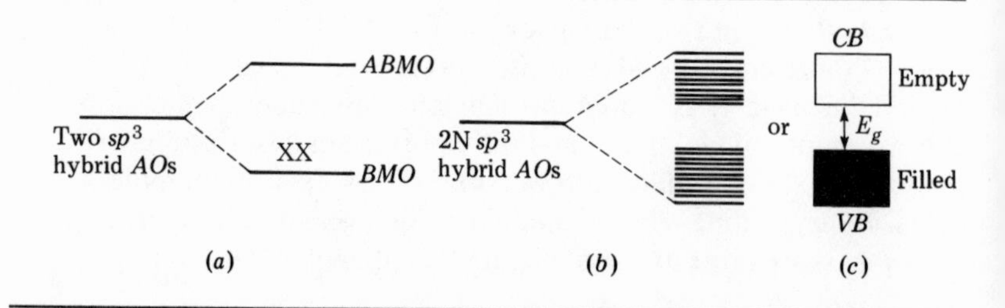

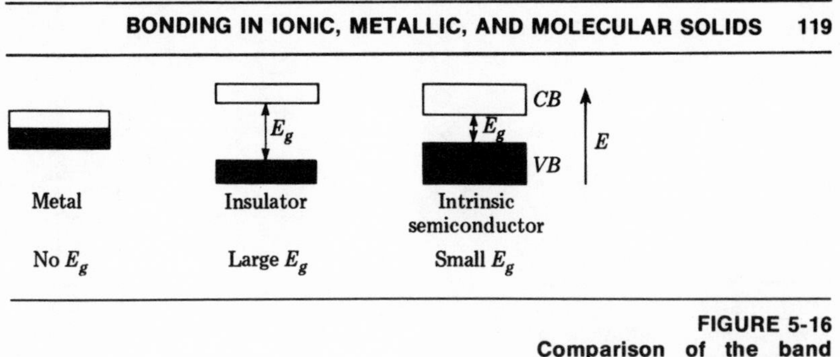

FIGURE 5-16
Comparison of the band schemes of a metal, an insulator, and an intrinsic semiconductor.

major difference is that, in a *metal*, the higher-energy empty states necessary for electrical conduction are immediately accessible to the electrons in the band; in an *insulator* no such states are accessible (barring dielectric breakdown under very high voltages), while in an intrinsic *semiconductor*, the energy gap is small enough that an increase in temperature of the solid gives some of the electrons in the valence band enough energy to cross into the empty upper band and move under the influence of a potential difference. As the temperature of the semiconductor increases conduction *increases* because more electrons become excited. (Recall that metallic conduction always *decreases* with increasing temperatures.)

Another class of semiconductors includes what are called *extrinsic* or impurity semiconductors. An intrinsic semiconductor like Ge with 1 ppm (1 in 10^6) atoms of As or Ga as a substitutional impurity shows a *thousandfold* increase in conductivity over pure germanium. This is a much more dramatic change than that observed in metals when impurities are present and is opposite in direction. (In metals, impurity atoms scatter the electron waves and decrease conductivity.)

To understand this effect consider the two-dimensional picture of an As atom in the tetrahedral Ge lattice shown in Fig. 5-17. The As atom replaces one Ge atom in the lattice and has itself one extra electron (s^2p^3) compared to all Ge atoms (s^2p^2). Within the band model these extra electrons occupy discrete

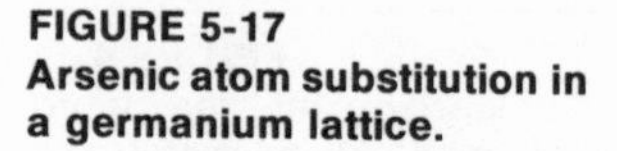

FIGURE 5-17
Arsenic atom substitution in a germanium lattice.

localized energy levels slightly below (about 0.8 kJ) the conduction band in Ge (Fig. 5-18*a*). These levels result from antibonding *MO*s between As and Ge orbitals, lower in energy than those in the conduction band because of the increased nuclear charge on As. Conduction in this *n-type* impurity semiconductor results when negative electrons are excited from these discrete levels to the conduction band. When impurity atoms with fewer electrons than the host are introduced into a crystal, for example, Ga (s^2p) into Ge (s^2p^2), a *p-type* semiconductor results (Fig. 5-18*b*). In this situation discrete empty levels exist right above the valence band. As electrons are excited into these states, positive holes are created in the valence band. Motion of these positive holes is responsible for conduction of electricity.

n- and p-type semiconductors in combination are used in the construction of *transistors*, which have replaced old bulky

FIGURE 5-18
Impurity semiconductors at O K.

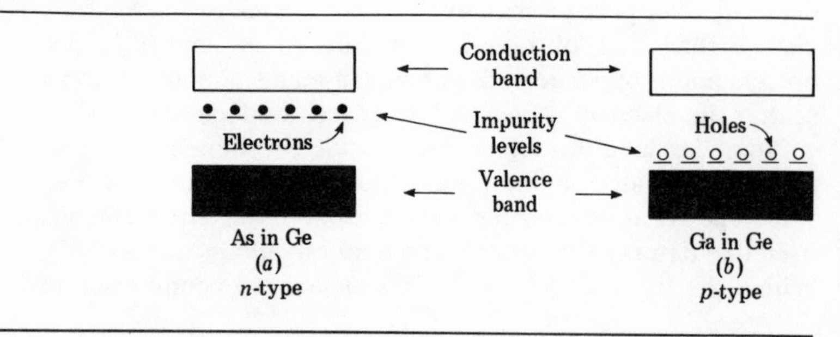

electron tubes and permitted design of electrical circuits small enough to be passed through the eye of a needle!

THE NATURE OF VAN DER WAALS FORCES

5-8 Of all the forces playing a role on the molecular stage those termed van der Waals forces are no doubt the weakest though probably the most universal. Associated with energies of about 0.4 to 40 kJ/mol, they are usually masked by stronger covalent forces in molecules, with typical energies of 400 kJ/mol. Their role is important only in explaining interactions between molecules and atoms with "saturated" orbitals, where no further covalent bonding is likely.

As early as 1873 Diderick van der Waals recognized the existence of weak attractive and repulsive forces among the molecules of a gas and attributed to them observed deviations from the ideal gas law: $PV = nRT$. Although his contribution to the elucidation of these forces was confined to correcting the gas law through empirically determined constants, his ideas spurred others on to the investigation of their nature.

Unlike covalent bonding, which is effective at small internuclear distances and is associated with electron overlap and exchange and consequently higher energies, van der Waals bonding may operate at distances where there is little or no overlap or exchange, and it is generally associated with smaller energies. For example, at about −160°C, Cl_2 gas crystallizes in a layer structure. The units are known to be covalent Cl_2 molecules with interatomic distances of 0.198 nm, almost exactly that of gaseous Cl_2. Van der Waals forces operate to hold the molecules in a layer (at distances from 0.33 to 0.38 nm) and also to hold the layers together (at distances from 0.37 to 0.39 nm). Solid Cl_2 sublimes to gaseous diatomic molecules when supplied a heat of sublimation of only 25 kJ/mol. In comparison the diatomic covalent bond energy is 238 kJ/mol.

There are at least four types of forces which contribute to the van der Waals bonding; we shall examine each separately.

The first force, attractive in type, leads to what is called the *orientation energy* and is present in molecular arrays whose constituent particles possess a permanent dipole moment, for

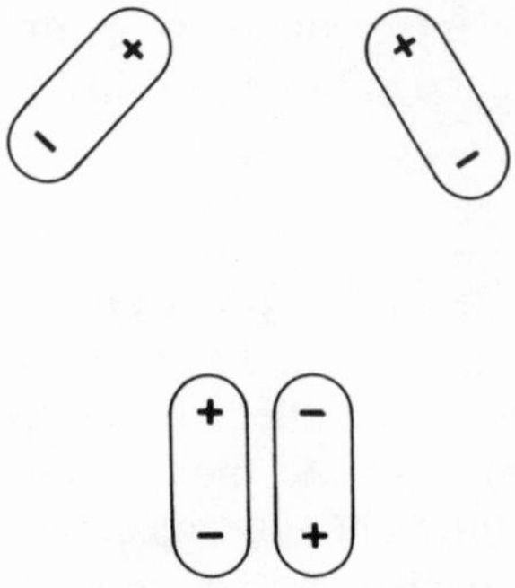

FIGURE 5-19
Dipole-dipole interaction. The plus and minus signs represent the centers of positive and negative charge for the resultant molecular dipole.

example, HCl, NH_3, H_2O. Consider the two dipoles shown in Fig. 5-19. Obviously their electrostatic interaction would be most attractive if they were aligned with positive end to negative end, as shown in the second part of the diagram. This desire of two dipoles to be perfectly oriented with respect to one another is felt at reasonably large distances, and the tugging at one dipole by another leads to the attractive *orientation*, or dipole-dipole–interaction, energy. As we shall see, this contribution to the total van der Waals energy is relatively small.

The second type of attractive force is that between a molecule with a permanent dipole and a molecule (or atom) without one. For simplicity we have pictured a dipole and a large spherical atom in Fig. 5-20. If the atom is polarizable, its electron cloud may distort toward the positive end of the dipole molecule, so that the centers of positive and negative charge in the atom no longer coincide and an *induced* atomic dipole is formed. The attractive interaction between the permanent dipole and the induced dipole leads to the second contribution

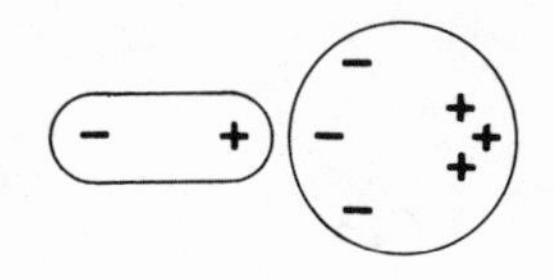

FIGURE 5-20
Dipole-induced dipole interaction.

to van der Waals binding, dubbed the *induction energy*, also a relatively small contribution.

Both the orientation energy and the induction energy can be adequately explained by classical physics. However, neither quantitatively accounts for the total van der Waals attraction in many compounds, and in particular neither accounts for the attractive forces responsible for the trend in boiling points of the noble gases, none of which possesses a permanent dipole moment. These boiling points are listed in Table 5-4.

The third kind of attractive force, leading to the *dispersion energy*, accounts for the noble-gas behavior and at the same time turns out to be usually the largest contributor to van der Waals attraction. Although the quantum-mechanical description gives a more accurate quantitative explanation, the simple picture afforded by classical theory is nice; hence we begin

TABLE 5-4
Boiling Points of Noble Gases in Degrees Kelvin

He	4.2	Kr	121
Ne	27	Xe	164
Ar	87	Rn	211

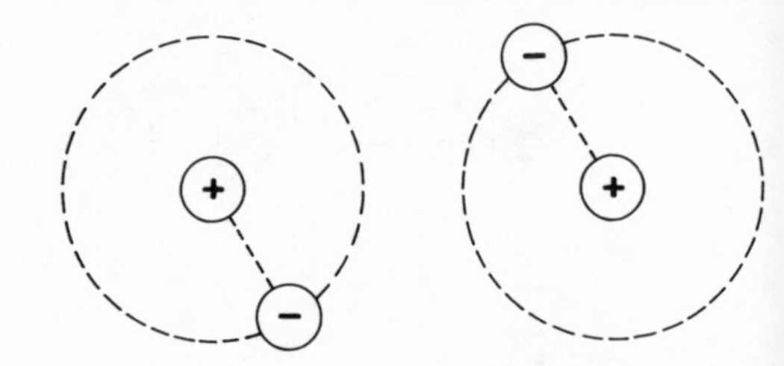

FIGURE 5-21
Classical dispersion interaction.

with that. We examine two hydrogen atoms at a sufficiently large distance from one another that their spherical clouds do not overlap appreciably. We then switch to the particle concept of the electron and consider a picture of the atom at some particular instant (Fig. 5-21). It is a so-called *instantaneous dipole*, which may induce in the neighboring H atom another instantaneous dipole, which fluctuates in phase with the first as the electrons move around the nuclei. The dispersion energy was attributed to the attraction between these two dipoles. However, the picture provided by this theory becomes rather complicated when we try to apply it to one of the larger noble gases.

The quantum-mechanical treatment offered by London in 1930 provides more insight into dispersion forces but is considerably more complicated. Here we say that two atoms (H atoms, for simplicity, though they are poor examples) even at distances precluding overlap "perturb" one another. In Fig. 5-22, electron cloud I feels attracted to nucleus *B*, cloud II to nucleus *A;* clouds I and II repel one another, nuclei *A* and *B* repel. The net result of all these interactions is an attractive energy, small compared to that of a covalent bond (where electron clouds merge) but large compared to orientation and induction energies.

The fourth force, necessarily a large and repulsive one, becomes effective when filled electron clouds on the interacting atoms or molecules begin to overlap. It is intimately related to the Pauli exclusion principle and is the same force which in

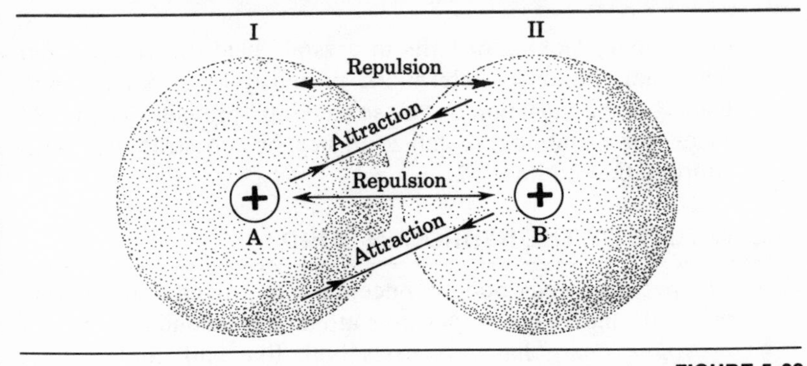

FIGURE 5-22
London dispersion interaction.

ionic crystals balances the electrostatic attraction at the equilibrium interionic distance.

Table 5-5 shows the magnitudes of orientation, induction, and London dispersion energies calculated for a few small molecules. Note that orientation energies increase as the molecular dipole increases, for example, in the series HI, HBr, HCl. In most cases the London dispersion energies predominate and for similar molecules (HCl, HBr, HI) *increase with the number of electrons* in the molecule. This latter point is one of the most important conclusions drawn from the quantum-mechanical theory.

Unlike covalent bonding, van der Waals binding is non-

TABLE 5-5
Contributions to the van der Waals Attractive Energy of Molecules at 293 K and a Separation of 0.4 nm

MOLECULE	PERMANENT DIPOLE MOMENT, *D*	ORIENTATION ENERGY, kJ/MOL	INDUCTION ENERGY, kJ/MOL	LONDON DISPERSION ENERGY, kJ/MOL
HI	0.38	0.005	0.025	5.62
HBr	0.78	0.091	0.060	2.59
HCl	1.03	0.274	0.079	1.54
CO	0.12	0.00005	0.0008	0.99
NH_3	1.5	1.24	0.15	1.37
H_2O	1.84	2.79	0.15	0.69

directional. Packing of units in crystals held together by van der Waals bonds is determined primarily by geometry, and usually some variation of "close packing" prevails. The noble gases, for example, crystallize in hcp or ccp structures with coordination number 12.

THE NATURE OF THE HYDROGEN BOND

5-9 The peculiar weak bond connecting a hydrogen atom on one molecule and an electronegative atom on a second molecule, a *hydrogen bond*, has properties both like and unlike ionic, covalent, or van der Waals bonds. The energies associated with hydrogen bonds are like those of strong van der Waals interactions (0.4 to 40 kJ/mol); however, in contrast to the nondirectional van der Waals bonds, hydrogen bonds do form in preferred directions. A hydrogen bond occurs between polar covalent molecules but is (according to one model) itself electrostatic in nature. It can be either *inter*molecular or *intra*molecular.

Some of the experimental evidence for hydrogen bonding is illustrated in the trends in the boiling points of the CH_4, NH_3, H_2O, and HF families shown in Table 5-6. Note that in this table, all molecules in the same row are isoelectronic, for example, SiH_4, PH_3, H_2S, and HCl of the second row. Consequently, if the forces between the molecules in the liquids were solely of the van der Waals type, we would expect parallel gradual increases in boiling points down the columns as the family members increase their number of electrons. This *is* true for the methane family.

TABLE 5-6
Boiling Points in Degrees Centigrade of Neighboring Hydride Families

NUMBER ELECTRONS	METHANE FAMILY		AMMONIA FAMILY		WATER FAMILY		HF FAMILY	
10	CH_4	−164	NH_3	−33	H_2O	+100	HF	+20
18	SiH_4	−112	PH_3	−87	H_2S	− 61	HCl	−85
36	GeH_4	− 90	AsH_3	−55	H_2Se	− 41	HBr	−67
54	SnH_4	− 52	SbH_3	−18	H_2Te	− 2	HI	−35

However, in the NH_3, H_2O, and HF families the lightest molecules have unusually high boiling points, hinting of *additional* intermolecular forces in action. The extra forces are characteristic of molecules containing H atoms covalently bonded to a very electronegative atom such as F, N, or O. In H_2O or in liquid HF the oxygen or fluorine atom has the lion's share of the electron pair involved in the σ bond. If we recall a previous picture, in a polar HX bond (where X is O, N, or F) the hydrogen atom, with a relatively small share of the electron pair, is almost a bare proton sitting on the end of the H—X sigma *MO* cloud. The lone-pair electrons of element X on a nearby molecule bind themselves through electrostatic attraction to this highly positive proton. The resulting "bond," in this case *inter*molecular, is a hydrogen bond.

In a second popular model of the hydrogen bond, a three-center molecular orbital is formed between the bond-pair orbital on one water molecule, the 1*s* orbital on hydrogen, and a lone-pair orbital on a second water molecule. (Return to Fig. 4-31, substituting a bond pair and a lone pair for the fluorine *p* orbitals and a 1*s* orbital for the Xe orbital.) The four electrons are assigned to the *BMO* and the *NBMO*. With no antibonding electrons the result is a *three-center* bond stabilized by two bonding electrons.

Because of the very small size of H^+ (remember that is has *no* electrons whatsoever), it can accommodate only 2 electron-pair clouds near it at one time. Thus the largest "coordination number" of H in a hydrogen bond is 2.

In liquid and solid HF, zigzag chainlike species occur, where the solid (shorter) lines represent covalent bonds and the dotted (longer) ones hydrogen bonds.

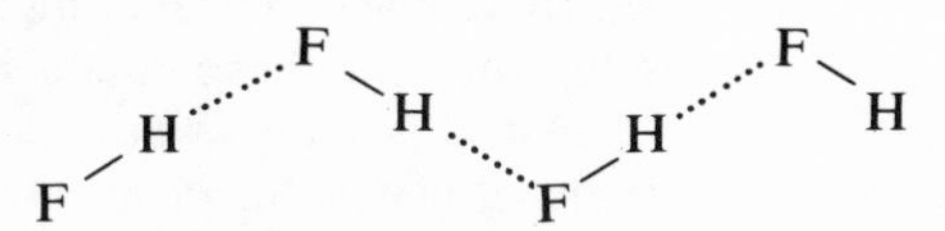

Note the linearity of the [X—H · · · · X] system. Presumably the electrostatic repulsion between the bond-pair clouds surrounding the H atom is minimized by this linear arrangement. Figure 5-23 illustrates this principle more clearly. Water molecules

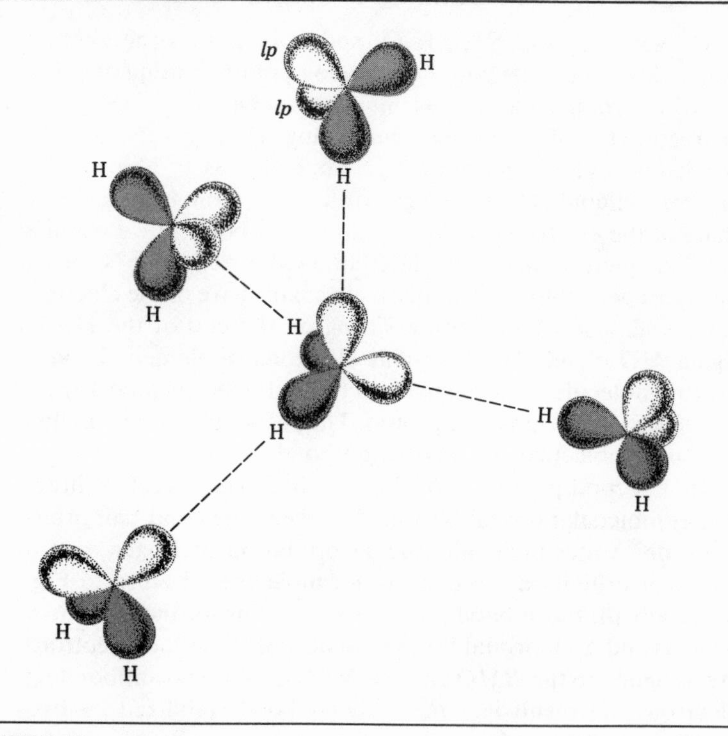

FIGURE 5-23
Hydrogen bonding and tetrahedral association in water.

are known to be tetrahedrally associated in ice, an association only partially broken down in liquid water. The darker lobes of the sp^3 hybrids are meant to be bond pairs, the lighter lobes lone pairs. Note the linearity of the *lp*-H-*bp* group.

The extra energy needed to break these bonds when liquid water is converted to isolated gaseous water molecules is reflected in the boiling point of water. Species containing central atoms which are not particularly electron-greedy (C, S, P) do not exhibit appreciable hydrogen bonding.

*Intra*molecular hydrogen bonds are known in molecules containing neighboring —XH and —Y appendages, for example, *ortho*-fluorophenol

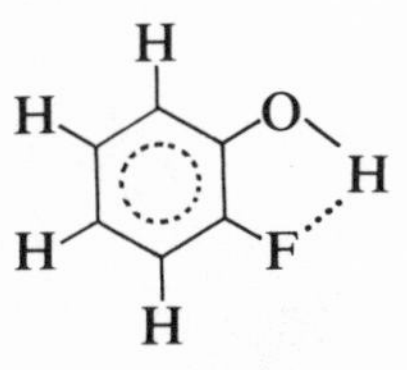

where the dashed circle represents the delocalized π ring of the benzene framework and the dotted line shows hydrogen bonding between the —OH group (which makes benzene a phenol) and the nearby fluorine atom. In this case the O-H-F system cannot be collinear and is thus associated with a weak hydrogen bond.

SILICATES AND GLASSES

5-10 In probably no other class of compounds is the internal electronic structure so clearly reflected in the macroscopic bulk properties of the materials as in silicates and glasses. We will examine the properties of these compounds as an application of orbital theory.

The basic unit of all silicates is a tetrahedral SiO_4 group of atoms with the silicon atom in the center. Depending on *how* these units are joined together in a solid and what other atoms are present, the material may be a gem, glass, tile, cement, asbestos, clay, vermiculite, mica, or quartz.

To see how these materials are formed we will first examine the electronic structure of the SiO_4 group. Like carbon, the silicon atom, with its $3s^23p^2$ outer configuration, goes into an sp^3 tetrahedral hybrid state with one electron in each hybrid. Each oxygen ($1s^22s^22p_x^22p_y^12p_z^1$) uses *one* of its unpaired electrons in a covalent σ bond with silicon; it can pair the other electron either by accepting an electron from some other atom, thus producing an ion, or by forming a second covalent bond with another silicon atom. By the latter approach one-dimensional chains, two-dimensional sheets, or three-dimensional networks may be built, with linking oxygen atoms shared by two silicon atoms.

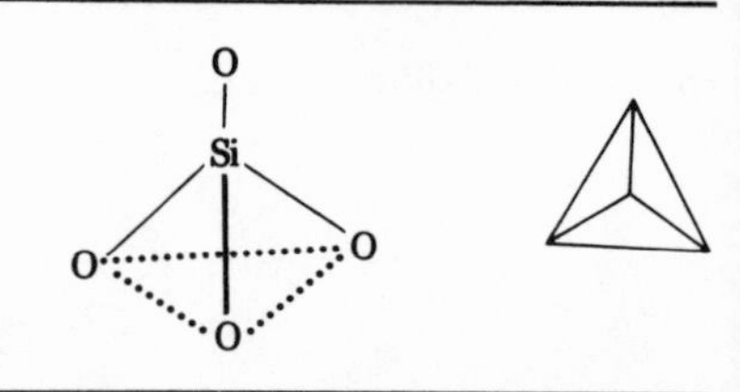

FIGURE 5-24
The SiO_4^{4-} ion.

The two extreme structures of silicates are represented by crystalline silicon dioxide, SiO_2, in which *every* oxygen atom forms two covalent bonds, and the orthosilicates, containing discrete SiO_4^{4-} ions, in which each oxygen forms only one covalent bond. Starting with discrete ion compounds, we will survey the structures and properties of a few important materials built from SiO_4 tetrahedra.

DISCRETE SILICATE ION STRUCTURES The structure of the SiO_4^{4-} ion is shown in Fig. 5-24, along with a convenient "shorthand" diagram representing the tetrahedral unit. The Si atom is at the body center with oxygen atoms at each apex, and the tetrahedron is viewed from the top. In the SiO_4^{4-} ion each oxygen after bonding to silicon has paired its unpaired electron by accepting one from a metal atom of the compound, thus giving the unit a total charge of −4. Portland cement, for example, Ca_2SiO_4, contains Ca^{2+} ions and SiO_4^{4-} ions held together by ionic bonds.

The $Si_2O_7^{6-}$ ion shown in Fig. 5-25 illustrates the first step

FIGURE 5-25
The $Si_2O_7^{6-}$ ion.

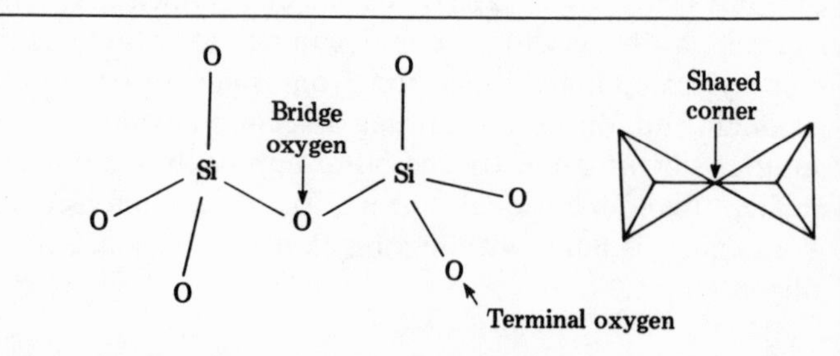

in building complex silicate structures. Note that while there are two kinds of oxygen atoms, bridging and terminal, there is only *one* kind of silicon atom.

In determining the formula and charge we may assume that terminal oxygen atoms have one minus charge, while bridging oxygens are neutral.

In joined tetrahedral units the bridging Si—O—Si bond angles are not 90°, but, depending on the compound, vary from 133 to 180°, indicating variable hybridization on the oxygen atom. The O—Si—O angles are *always* 109.5°.

CHAIN STRUCTURES Chains of silicate groups occur commonly in two forms, the pyroxene single chain or the amphibole double chain, Fig. 5-26. These chains have as-

FIGURE 5-26
Chain structures. (*a*) Pyroxene chain; (*b*) amphibole chain.

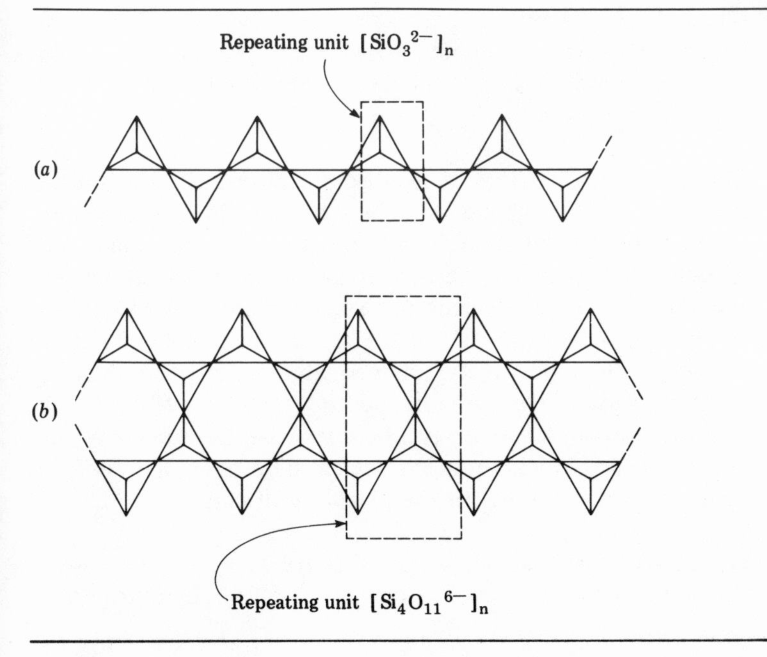

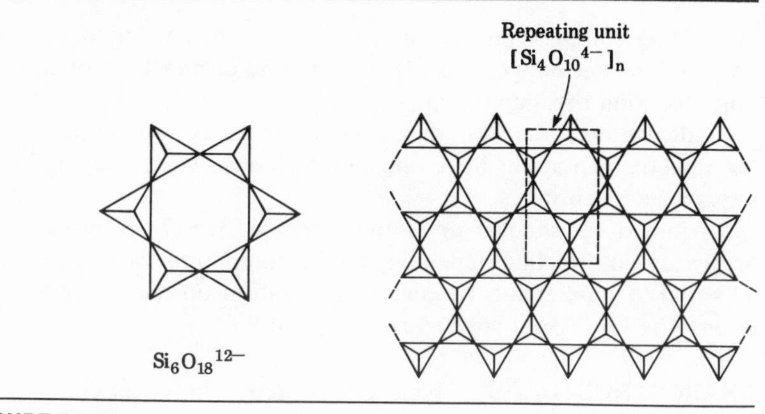

FIGURE 5-27
Sheet structures.

sociated with them a resultant negative electrical charge distributed among the terminal oxygen atoms along the chain. In solids like *asbestos* metal ions located in large holes in the structure bind the chains together loosely by electrostatic attraction. The fibrous, stringy nature of asbestos (which contains both single and double chains) results from the weakness of these *inter*chain bonds relative to the strong bonds within each silicate chain.

SHEET STRUCTURES The discrete planar $Si_6O_{18}^{12-}$ ion of Fig. 5-27 with Be^{2+} and Al^{3+} ions makes the gem emerald, $Be_3Al_2Si_6O_{18}$. Continued in all directions in the plane, interlocked hexagons may form sheets with one terminal oxygen and three bridging oxygens per silicon atom (Fig. 5-27). In clays, like kaolinite, $Al_4(OH)_8Si_4O_{10}$, the silicate sheet is meshed with a positively charged aluminate $[Al_4(OH)_8^{4+}]_n$ sheet to make an electrically neutral sandwich (Fig. 5-28). Since an array of sandwiches is held together by van der Waals forces, the kaolinite sandwich sheets slide over one another easily, thus accounting for the plasticity of clay.

THREE-DIMENSIONAL STRUCTURES A three-dimensional network structure is found in a high-temperature crys-

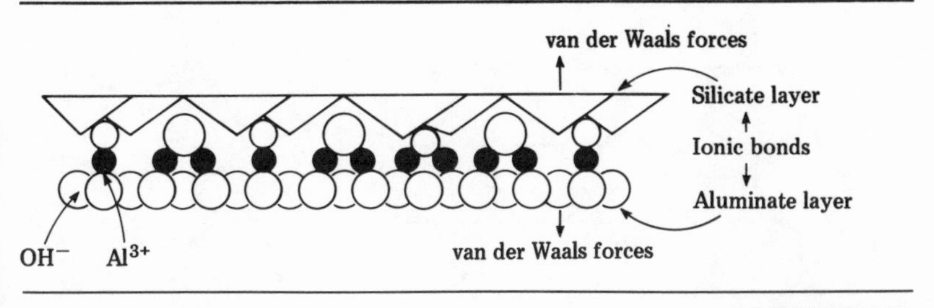

FIGURE 5-28
Kaolinite sandwich.

talline form of quartz SiO_2 called cristobalite (Fig. 5-29). Here *all* the oxygen atoms are shared by two silicon atoms, all bonds are covalent, and the structure has a characteristic long-range order. Silicon atoms are located at the corners, face centers, and inside of the unit cell, with oxygen atoms between each silicon pair. Each unit cell contains rings with six silicon atoms and six oxygen atoms.

In contrast, the structure of silica *glass*, still three dimensional, is characterized by short-range order. Glass forms when molten silica SiO_2 is cooled rapidly to a rigid state before all the chains have a chance to take their proper ordered positions. A two-dimensional comparison of crystal and glass structures is shown in Fig. 5-30.

Glasses, unlike ordered crystalline solids, do not have clear-cut melting points, but rather "softening" points, temperatures at which the rigid structures begin to loosen up. The softening point of silica glass is lower than the melting point of quartz,

FIGURE 5-29
Cristobalite quartz.

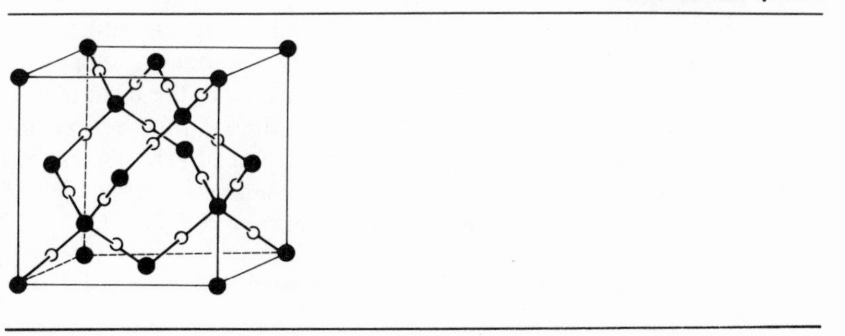

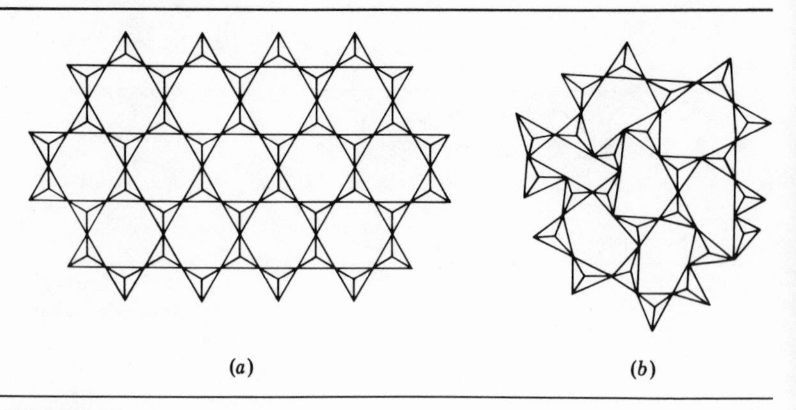

FIGURE 5-30
Crystal (*a*) versus glass (*b*) structures.

perhaps because of the strain in the twisted, distorted rings. Silica glass has a low thermal-expansion coefficient; i.e., it does not expand much with increase in temperature (a valuable property, since we don't want window panes to expand in the summer or contract in the winter). However, as a construction material, pure silica when softened is very viscous and difficult to work with. Bringing chemistry into action, we add fluxing species to the molten silica in the form of *network breakers* or *network formers* (Fig. 5-31). Network breakers are Na^+ or Ca^{2+} ions, which break chains, neutralize the resulting terminal O^- ions, and reduce the viscosity. The result is soda glass, or the soft glass characteristic of most bottles. A typical network former is boron, which replaces some of the three-dimensional crosslinks with two-dimensional trigonal bond groups. The resulting borosilicate (marketed as Pyrex or Kimax) is more viscous than soda glass but has a higher softening point.

Molecular science—studies of orbitals, bonds, shapes of molecules, and structures of solids—has led to the birth of molecular engineering, where, as demonstrated in this section, abstract theories developed by chemists and physicists may be put to use by engineers in the understanding of properties of old materials and the design of new and better ones.†

†A paraphrase on an articulate and poetic statement by von Hippel, Ref. 2.

EXERCISES

1 Formulate an expression for the stabilization energy, E_{ic}, when four Na^+ and four Cl^- ions form an ion cube of side d. Then compute the total stabilization energy, U_{ic}, resulting when 1 mol of Na^+ ions and 1 mol of Cl^- ions come together as cubes. Show that the results are intermediate between ion-square and lattice energies.

2 The ionization energy of Mg ($I_1 + I_2$) is 2179 kJ/mol. To form 0^{2-}, 694 kJ/mol must be supplied to O. Assuming an

FIGURE 5-31
Network breakers (*a*) and network formers (*b*).

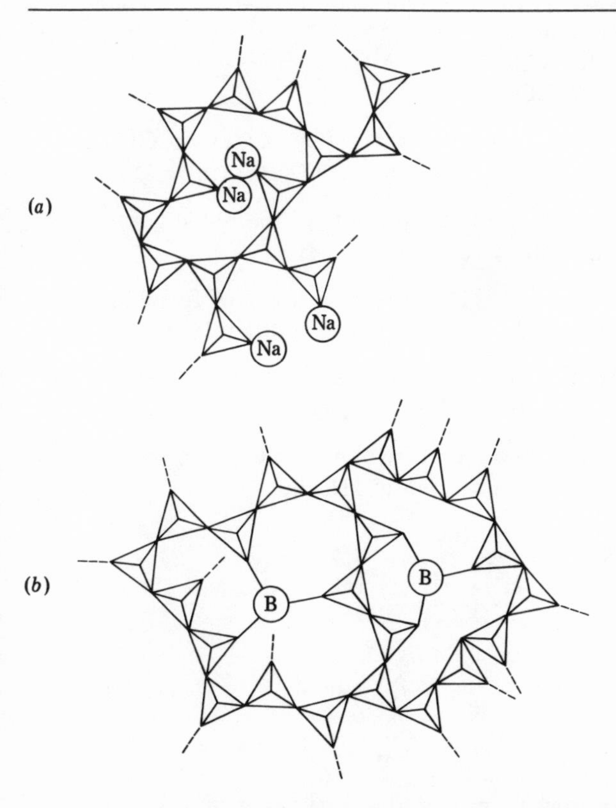

NaCl-type structure, compute the stabilization energy of solid MgO relative to the isolated gaseous atoms.

3 A correction to the electrostatic model for lattice energies which accounts for repulsion of electron clouds on neighboring ions has the general form

$$E_{\text{rep}} = \frac{N A_M}{n} \times E_{ip}$$

where E_{ip} is the energy of an ion pair and n is a parameter evaluated from the compressibility of the solid. For NaCl, $n = 9.1$. Compute this correction to the lattice energy given in Fig. 5-1, and compare your result to the experimental lattice energy of NaCl, 757 kJ/mol.

4 Using the ionic potential concept, select the most ionic and the least ionic compound in each of the following groups:

(*a*) $BiCl_3$, $ScCl_3$, $BeCl_2$, $AlCl_3$
(*b*) TiO_2, CeO_2, SiO_2

5 Predict the coordination number of Ce^{4+} in CeO_2; Ca^{2+} in CaF_2; Ti^{4+} in TiO_2; Mg^{2+} in MgTe; Ba^{2+} in BaO.

6 Explain why the boiling points of the noble gases (Table 5-4) increase with increase in atom size.

7 KF reacts with HF to form a solid of formula KHF_2. Discuss the probable geometry and bonding in the HF_2^- ion.

8 Discuss the probable relative extent of hydrogen bonding in CH_4, CH_3OH, CH_3F, CH_3NH_2.

9 One might think that two possible types of sideways branching could occur in liquid HF, i.e.,

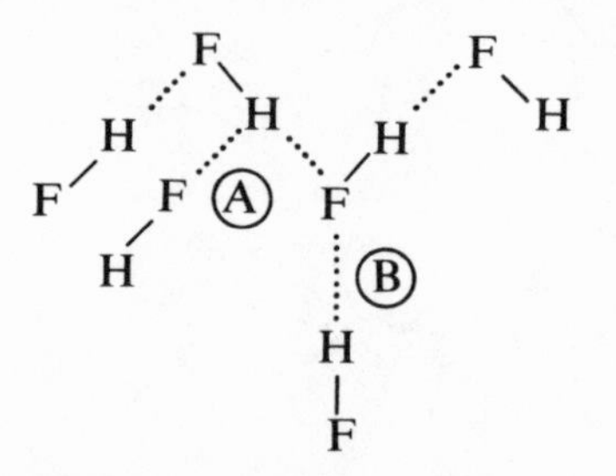

Discuss the relative probability of types A and B.

10 The density of liquid water is considerably greater than that of solid water; i.e., ice floats on water. Suggest an explanation in terms of hydrogen bonding and structure.

11 At room temperature F_2 and Cl_2 are gases, Br_2 is a liquid, and I_2 is a solid. Why?

12 Planar boric acid molecules, $B(OH)_3$, crystallize in the sheet structure illustrated below. Boron atoms are at trigonal sites and OH groups at terminal positions.

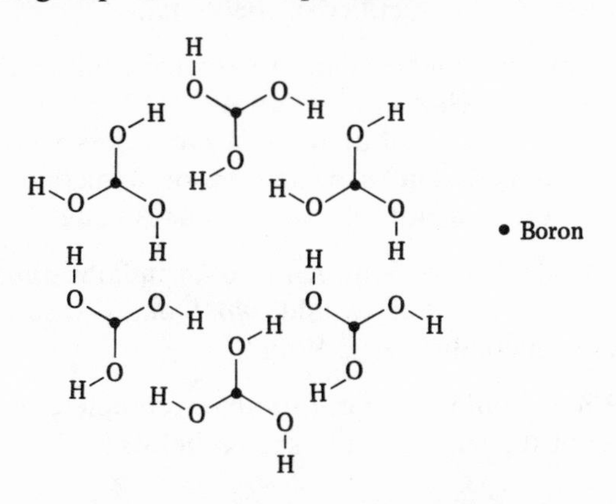

It is more difficult to break a sheet than to separate one sheet from another. What kind of bonding holds the molecules together within a sheet? Complete the above picture by sketching in these bonds. What kind of bonding occurs *between* layers?

13 LiH and LiF are crystalline salts with the same structure as NaCl. The H^- radius is 0.208 nm while that of F^- is 0.136 nm. Which salt would you expect to have the higher melting point, LiH or LiF? Why?

14 Without consulting a table of ionic radii, decide which is the largest ion in each group below:

(*a*) O^{2-}, S^{2-}, F^-, Cl^-
(*b*) Na^+, Mg^{2+}, O^{2-}, F^-

15 The metals below all have a bcc crystal structure. Which would have the lowest melting point?

Li Na K Rb Cs

16 An alloy of 0.5% C in Fe would be a (better, poorer) conductor of electricity than pure iron. Which would most likely support a heavier load: a wire of pure iron or a wire (of same diameter) of the above alloy? Why?

17 When a metal or alloy is "cold-worked," i.e., hammered at low temperatures, there is a large increase in the number of defects (particularly dislocations) in the structure. Predict the effect of cold-working on (*a*) the electrical conductivity of the metal and (*b*) the ductility of the metal.

18 Many compounds that are isoelectronic with Group IV solids, such as GaAs, have tetrahedral lattices and are semiconductors. If a small amount of Ge atoms were substituted for Ga atoms in GaAs what type of semiconduction would occur? If Ge substitution occurred at As sites?

19 Introduction of S atoms into an indium phosphide (InP) semiconductor lattice would *most likely* result in (*n*-type, *p*-type) semiconductivity. Why?

20 What would be the formulas (complete with negative charge) of the silicate units shown below?

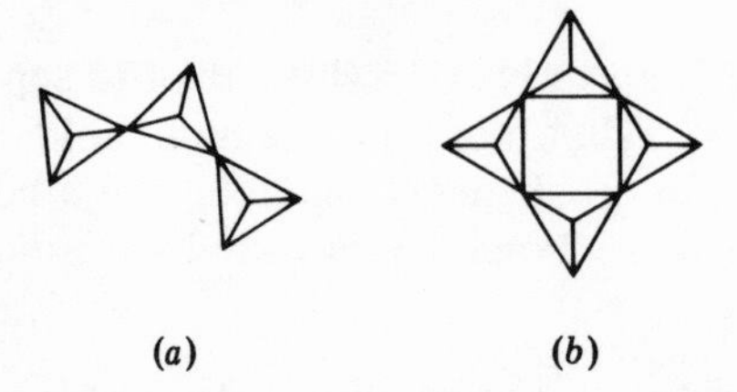

(*a*) (*b*)

21 The effect of adding $MgCO_3$ to molten silica SiO_2 would include which of the following? (More than one answer is possible.)

(*a*) Bubbles of CO_2 would form.

(*b*) Silicate chains would be broken.

(*c*) The molten silica would become more viscous.

(*d*) The resulting glass would be an excellent conductor of electricity.

(*e*) The resulting glass would have a softening point lower than that of pure silica.

22 Predict the probable internal structure of the minerals:
(*a*) Talc $Mg_3(OH)_2Si_4O_{10}$
(*b*) Diopside $CaMg(SiO_3)_2$
(*c*) Zircon $ZrSiO_4$

REFERENCES FOR FURTHER READING

GENERAL

1 Coulson, C. A., "Valence," 2d ed., chaps. 12, 13, Oxford University Press, New York, 1961.

2 Von Hippel, A. R., Molecular Designing of Materials, *Science*, **138,** 91 (1962).

3 Heslop, R. B., and P. L. Robinson, "Inorganic Chemistry," 3d ed., chaps. 5, 9, Elsevier Publishing Company, Amsterdam, 1967.

4 "Materials," A Scientific American Book, W. H. Freeman and Co., San Francisco, 1967.

IONIC BONDING AND CRYSTAL STRUCTURE

5 Gehman, W. G., Standard Ionic Crystal Structures, *J. Chem. Educ.*, **40,** 53 (1963).

6 Sime, R. J., Some Models of Close Packing, *J. Chem. Educ.*, **40,** 61 (1963).

7 Ho, S.-M., and Bodie E. Douglas, A Broader View of Close Packing to Include Body-centered and Simple Cubic Systems, *J. Chem. Educ.*, **45,** 474 (1968).

8 Ho, S.-M., and Bodie E. Douglas, A System of Notation and Classification for Typical Close-packed Structures, *J. Chem. Educ.*, **46,** 207 (1969).

9 House, J. E., Jr., Ionic Bonding in Solids, *Chemistry*, **43,** 18, February (1970).

10 Burrows, E. L., and S. F. A. Kettle, Madelung Constants and Other Lattice Sums, *J. Chem. Educ.*, **52,** 58 (1975).

11 Quane, Denis, Crystal Lattice Energy and the Madelung Constant, *J. Chem. Educ.*, **47,** 396 (1970).

12 Evans, R. C., "An Introduction to Crystal Chemistry," 2d ed., Cambridge University Press, London, 1966.

13 Sanderson, R. T., The Nature of "Ionic" Solids, *J. Chem. Educ.*, **44,** 516 (1967).

METALS AND SEMICONDUCTORS

14 Lefever, R. A., An Introduction to the Electron Theory of Metals, *J. Chem. Educ.*, **30,** 486 (1953).

15 Lefever, R. A., A Summary of Semiconductor and Transistor Theory, *J. Chem. Educ.*, **30,** 554 (1953).

16 Weller, Paul F., An Introduction to Principles of the Solid State: Extrinsic Semiconductors, *J. Chem. Educ.*, **48,** 831 (1971).

17 Gurnee, Edward F., Fundamental Principles of Semiconductors, *J. Chem. Educ.*, **46,** 80 (1969).

18 Young, J. A., and J. G. Malik, Chemical Queries (Colors of Metals), *J. Chem. Educ.*, **48,** 270 (1971).

19 Keller, E., Transistor Effects—Panacea or Pandora's Box?, *Chemistry*, **47,** 7, September (1974).

20 Verhock, F. H., What is a Metal?, *Chemistry*, **37,** 6, November (1964).

21 Van Reuth, E. C., An Analogy for the Band Theory of Metals, *J. Chem. Educ.*, **43,** 484 (1966).

22 Weller, Paul F., An Introduction to Principles of the Solid State, *J. Chem. Educ.*, **47,** 501 (1970).

23 Weller, Paul F., An Analogy for Elementary Band Theory Concepts in Solids, *J. Chem. Educ.*, **44,** 391 (1967).

24 Beveridge, David L., and Bernard J. Bulkin, Descriptive Crystal Orbital Theory of Conduction in Diamond and Graphite, *J. Chem. Educ.*, **48,** 587 (1971).

INTERMOLECULAR BONDING

25 House, J. E., Jr., Weak Intermolecular Interactions, *Chemistry*, **45,** 13, April (1972).

26 Choppin, G. R., Water—H_2O or $H_{180}O_{90}$?, *Chemistry*, **38,** 7, March (1965).

27 McClellan, A. L., The Significance of Hydrogen Bonds in Biological Structures, *J. Chem. Educ.*, **44,** 547 (1967).

SILICATES AND GLASSES

28 Johari, Gyan P., Introduction to the Glassy State in the Undergraduate Curriculum, *J. Chem. Educ.*, **51,** 23 (1974).

29 Hicks, John F. G., Glass Formation and Crystal Structure, *J. Chem. Educ.*, **51,** 28 (1974).

30 Slabaugh, W. H., Clay Colloids, *Chemistry*, **43,** 8, April (1970).

31 Companion, A., and K. Schug, Ceramics and Glass, *Chemistry*, **46,** 23, October (1973).

STRUCTURE OF TRANSITION-METAL COMPOUNDS

6

INTRODUCTION

6-1

Chemists have long been fascinated by the transition metals, and for good reason: their compounds abound with intriguing magnetic properties, colors, and geometries. For example, consider the wide variety of color exhibited by just the transition-metal oxides: purple Ti_2O_3, bronze TiO, yellow V_2O_5, blue VO_2, red CrO_3, emerald green Cr_2O_3, brick red Fe_2O_3, pale green NiO, and olive MnO. Oxides of most nontransition metals are an unexciting black or white.

Since the early formulation of the orbital theory of atoms and ions, the *d* electrons on the transition metals have been considered responsible for these unusual properties. During the past twenty years a special version of orbital theory, crystal field theory, has been used extensively to correlate electronic structure, color, geometry, magnetic effects, and many other physical and chemical properties of transition-metal compounds. Crystal field theory has proved to be one of the most profitable and interesting applications of wave mechanics to chemical problems.

Before discussing the theory itself, we must digress briefly to consider the measurement and structural significance of color and magnetic properties.

COLOR

6-2 In Chap. 1 we learned that light in its particle nature consists of photons, each characterized by a wavelength or color, the wavelength being intimately related to the energy of the photon through the relationship

$$E = \frac{hc}{\lambda}$$

Only those photons whose wavelengths fall in the range 400–700 nm stimulate the human sense of sight, i.e., are visible. Red, the long-wavelength color, corresponds to low energy, and violet, with short wavelength, to high energy. Between these limits, with increasing energy, are photons associated with the colors orange, yellow, green, and blue. Ordinary white light, such as that emitted from a hot tungsten filament, is a continuum of all colors and energies, and black corresponds to absence of visibly detectable photons.

We see an opaque object primarily by light reflected from its surface. Consequently, an object illuminated by white light may appear red in color because it is absorbing high-energy photons and reflecting the lower-energy red ones from its surface. A simplified plot showing light absorption versus energy or color of the incident photon is shown in Fig. 6-1 for a red object, along with representative *absorption spectra* of some objects of other colors.

The mechanism of absorption of light by a solid or liquid is closely related to that of emission of photons by gaseous atoms excited in a flame (Chap. 2). When sodium atoms from a salt or salt solution are vaporized in a bunsen flame, electrons on the gaseous metal atoms are excited to higher energy levels and, as they return to the ground state, photons of energy corresponding to the color yellow are emitted. At room temperature, certain of the electrons in a solid or solution when bombarded with white light may absorb photons whose energies correspond either to allowed quantum jumps within an ion or to the energy necessary to transfer the electron from an outer orbital on one ion to an empty orbital on an ion of a different type. For example, the orange-yellow color of cadium sulfide

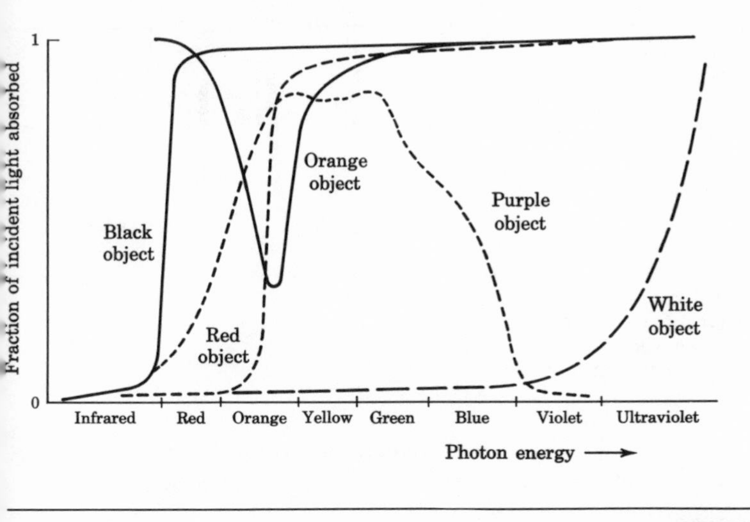

FIGURE 6-1
Absorption spectra.

is probably due to electron transfer of a $3p$ electron on a sulfide ion to a vacant orbital on a nearby cadmium ion. Those photons not acceptable for the process (the lower-energy red, yellow, and orange ones) are reflected back to us as the characteristic color of the compound. The transferred or excited electron usually dissipates its energy by contributing it to the vibrational motions of the ions in the solid lattice or to vibrational, rotational, and translational motions of molecules in solution. Consequently no emission process (light flash) usually occurs. Since the lifetime of the excited state is very short, the electron is soon back in its ground state ready for another absorption. Vibrational processes are not available to isolated atoms or monatomic ions in the gas phase; hence the existence of emission spectra for these.

MAGNETIC PROPERTIES

6-3 All substances with few exceptions contain atoms with one or more filled electron levels (s^2, p^6, d^{10}, . . .). When placed in the

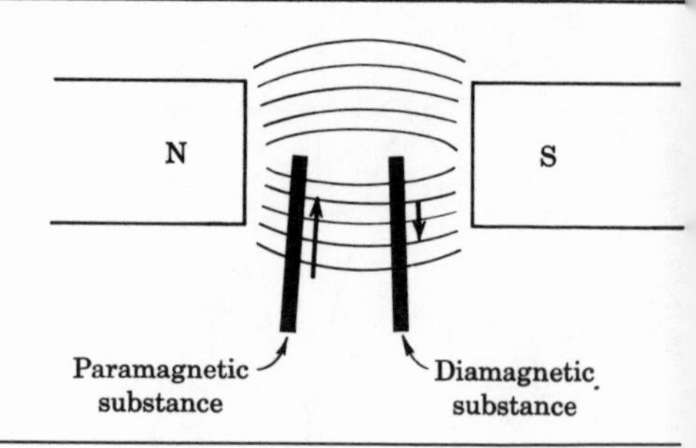

FIGURE 6-2
Behavior of paramagnetic and diamagnetic substances in a magnetic field.

magnetic field existing between the poles of a natural magnet or an electromagnet, such a substance experiences a weak repulsion by the field, tending to push it away. This phenomenon is known as *diamagnetism* and is independent of temperature (see Fig. 6-2).

In contrast, substances containing atoms with one or more unpaired electrons are strongly attracted into magnetic fields, an attraction inversely dependent on temperature. This phenomenon is known as *paramagnetism*. The forces associated with paramagnetism are considerably larger than those of diamagnetism and usually mask them almost completely. An atom with several completed levels and only one unpaired electron is still strongly paramagnetic.

The property known as *ferromagnetism* is rare, exhibited only by iron metal and several other substances which are attracted into magnetic fields by forces thousands of times larger than paramagnetic ones.

Though the theory of ferromagnetism is complex, we can get a qualitative picture of the causes of diamagnetism and paramagnetism through a simple model based on the laws of classical physics. When an electric current (a flow of negative charge) is passed through a circular loop such as that of Fig. 6-3, a magnetic field is induced perpendicular to the loop, with a north and south pole (like a little bar magnet) dependent upon the direction of electron flow. An electron undergoing orbital motion (think for a moment of a Bohr orbit) or spinning around its axis in an α or β way is a negative charge undergoing angular motion (though the loop in spin motion has zero radius), and as a result both spin and orbital motion have intrinsic magnetic fields associated with them. When two electrons are spin-paired, the magnetic fields are in opposite directions and cancel. When an atom or molecule has one or more *unpaired* electrons, it has a permanent resultant magnetic field associated with it, which in a sense causes it to act like a little bar magnet. When placed in an external magnetic field, the bar magnet will be attracted by forces tending to line up its poles with the poles of the field (north to south) insomuch as temperature permits. Increasing temperature increases the orientational motion of the bar magnets and makes alignment more difficult. Thus paramagnetic forces decrease with increasing temperature.

When electrons are part of a closed orbital level, not only their spin magnetic fields but also the fields associated with their orbital motion cancel (are "coupled"). When brought into the presence of an external field, the orbital "bar magnets" are uncoupled slightly, producing a resultant magnetic field which always *opposes* the direction of the external field, whatever it may be. In other words, if the diamagnetic substance is approaching the north pole of an external magnetic field, a resultant bar magnet with its north pole pointed at the field is created by uncoupling the orbital bar magnets. This is a repulsive situation, because like poles repel, and the substance is

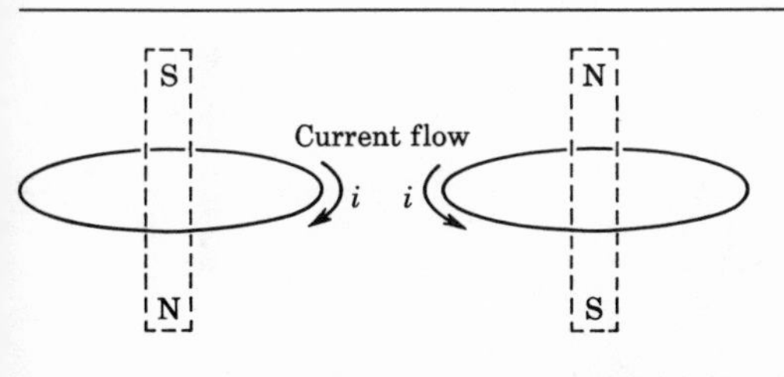

FIGURE 6-3
Bar magnet representation of the magnetic field induced by a flow of current through a loop.

pushed from the field. Unlike paramagnetism, diamagnetism is independent of temperature, because here there is no question of lining up rapidly moving molecules; the opposing magnetic field is created by the electrons no matter in which direction they approach the external field.

More details about these processes may be found in Ref. 2.

An apparatus for measuring these effects, the Gouy balance, is illustrated in Fig. 6-4. The sample to be studied, S, is hung from one arm of a sensitive analytical balance, so that it is hovering at the edge of the magnetic field produced between the poles of an electromagnet P. With the magnet turned off, the weight of the sample is balanced with known weights added to pan A. Then the electromagnet is turned on, producing a field of known strength, and the sample, if paramagnetic, undergoes an apparent increase in weight due to the pull of the field, or, if diamagnetic, a decrease in weight due to repulsion by the field. These weight changes can be measured by adding or removing weights from the pan until the balance pointer is again zeroed. Along with data on sample weight, composition, and temperature, these weight changes are used to compute a quantity called the magnetic dipole moment μ, analogous to the electric dipole moment described in Chap. 4. Usually μ is expressed as a multiple of a unit called the Bohr magneton

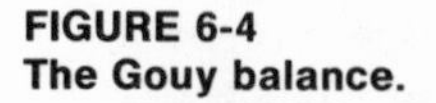

FIGURE 6-4
The Gouy balance.

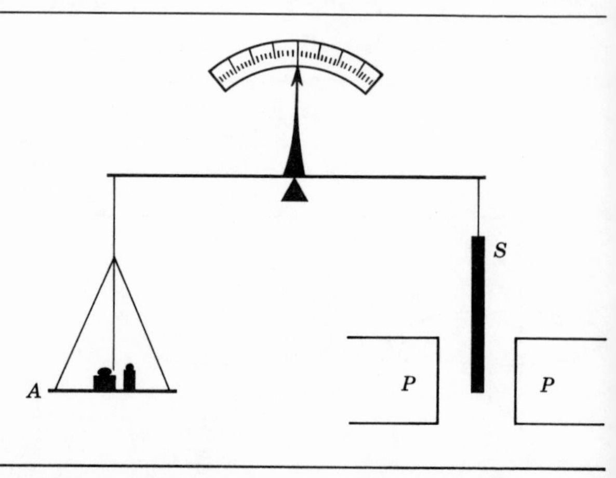

(BM). For an atom or molecule with n unpaired electrons the magnitude of μ is $[n(n+2)]^{\frac{1}{2}}$. We shall have occasion later to compute n from the measured magnetic dipole moments of transition-metal ions.

FUNDAMENTALS OF CRYSTAL FIELD THEORY

6-4 There are many excited energy levels available to electrons in isolated gaseous transition-metal ions, and the positions of these relative to the ground states are quite well known through careful study of emission spectra. Nevertheless these are not adequate to explain the absorption energies associated with the colors of transition-metal compounds. Any respectable theory of their structure must remedy this and must also be compatible with experimentally observed changes in color and magnetic properties with change in oxidation number and environment of the metal ion. Crystal field theory has succeeded in these and many other respects, and, fortunately, to understand its basic principles, we need accept only two things: the shapes of the $3d$ orbitals, as shown in Fig. 6-5, and the simple law of electrostatics that says like charges repel one another.

We may best explain the theory by considering an example, the perturbation of the $Ti^{3+}(3d^1)$ ion in an oxide lattice. We recall that in the isolated ion the single d electron has no preference as to which d orbital it will occupy, all five being degenerate with respect to energy and thus equally stable. Arbitrarily we put it in the central orbital in the energy-level diagram in Fig. 6-6. We then imagine a situation wherein 12 electrons in their cloudlike character are taken from some external pot of electrons and shaped into a spherical shell surrounding the Ti^{3+} ion at some known radius R, as illustrated in cross section in Fig. 6-7a. The single d electron on Ti^{3+} will find each orbital equally uncomfortable because of the spherical symmetry of the repulsive cloud. In terms of energy the system is less stable than the isolated ion (so far as the electron is concerned), but the orbital degeneracy is retained, as shown in Fig. 6-6. To bring our model closer to the real physical situation (six oxide ions arranged octahedrally around the Ti^{3+}), we

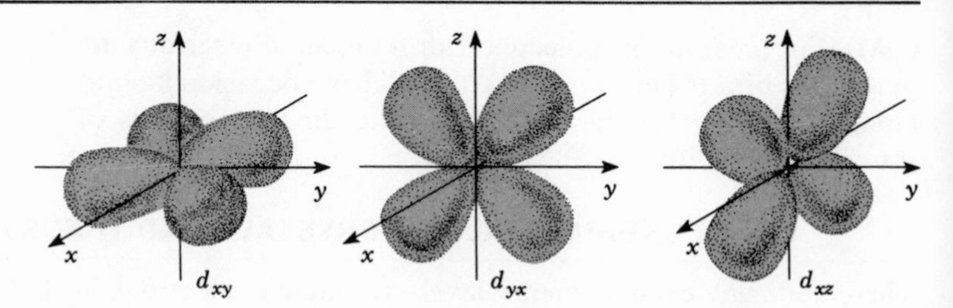

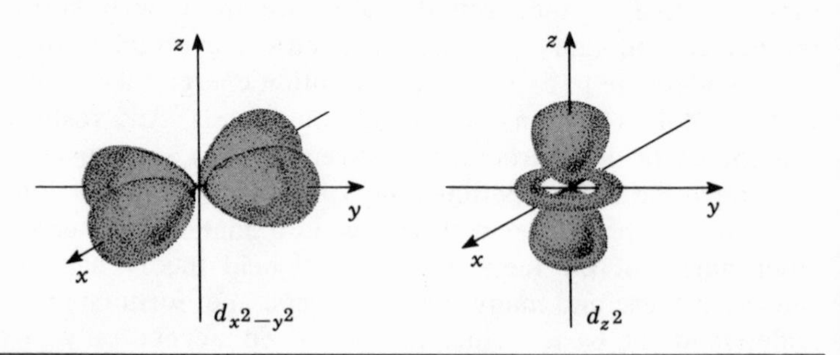

FIGURE 6-5
The five *d* orbitals.

FIGURE 6-6
Crystal field energy relationships for an octahedral system. Changes due to repulsion and attraction are qualitative, not quantitative.

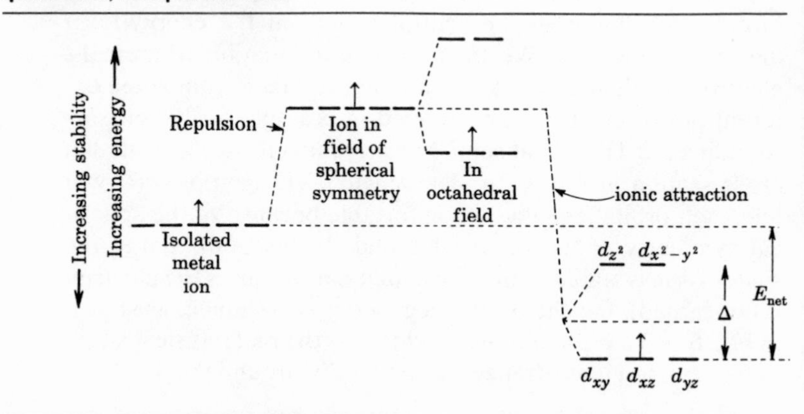

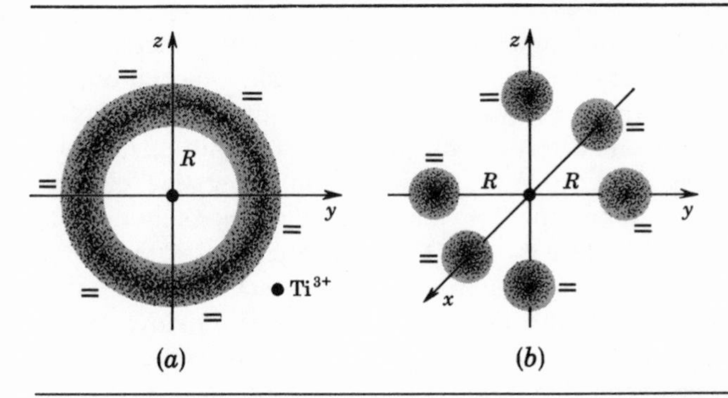

FIGURE 6-7
Formation of an octahedral field. (*a*) A spherical electron cloud surrounding the metal ion; (*b*) redistribution of charges to octahedral symmetry.

think of another process, in which the spherical cloud is rearranged so that two electrons are concentrated in very small volumes at distances $\pm R$ along each of the three coordinate axes in space (Fig. 6-7*b*). The single *d* electron now finds the d_{z^2} and the $d_{x^2-y^2}$ orbitals, with their lobes pointing directly at the clouds along the axes, electrostatically less favorable than in the spherically symmetric case, and the d_{xy}, d_{xz}, and d_{yz} orbitals, with lobes pointing into empty space between the axes, considerably more favorable. The *d* orbitals are no longer degenerate with respect to energy, and in fact are split into the two levels shown in Fig. 6-6. It is not obvious from the pictures that the d_{z^2} orbital is equivalent to $d_{x^2-y^2}$ in this environment, but wave-mechanical arguments (which we can only accept in this text) establish that it is so.† These treatments also show that, if the total energy difference between the upper and lower levels is designated Δ (Fig. 6-6), then the increase in stability and the decrease in stability over the spherically symmetric case are $\frac{2}{5}\Delta$ and $\frac{3}{5}\Delta$, respectively. This is simply a statement of energy balance: in our rearrangement of the

† For a description of five equivalent *d* orbitals, see Ref. 10. These pictures are less useful in our arguments than those of Fig. 6-5.

electron cloud the number of orbitals (3) which increased in stability multiplied by the increase $\frac{2}{5}\Delta$ equals the number of orbitals (2) which decreased in stability times the decrease $\frac{3}{5}\Delta$. This rule will always hold, whatever the splitting diagram.

We may now identify the six doubly negative electron clouds surrounding the Ti^{3+} ion with six O^{2-} ions in a crystal lattice of Ti_2O_3. Here the six O^{2-} are located around the Ti^{3+} in an arrangement closely approximating perfect octahedral symmetry. In this field the Ti^{3+} ions gains a *crystal field stabilization energy* (*CFSE*) of $\frac{2}{5}\Delta$, which we should not expect without knowledge of the shapes of *d* orbitals! The *CFSE*, as defined, is the difference between a hypothetical situation (where the oxide electrons are smeared into a spherical shell) and a real one (where they are localized near the oxygen nuclei) and is thus clearly not a directly measurable quantity. Nevertheless, as we shall see, the existence of crystal field stabilization energy (proved indirectly) strongly influences the properties of transition-metal compounds.

Before proceeding further, we must point out that the crystal field stabilization energy is a relatively small energy superimposed on the larger electrostatic energies present in an ionic crystal. To get the complete picture, we must recognize that besides the electronic repulsion energies described thus far the Ti^{3+} ion as a whole feels a strong electrostatic attraction for the six negative oxide ions. This is so large that, when it is combined with the electronic repulsion, the net energy of the system is lower than that of the isolated ions. This net energy is related to the classical crystal lattice energy U_L of Chap. 5 and is responsible for the stability of the crystal. The crystal field stabilization energy is a small contribution to this (see Fig. 6-6).

EXPLANATION AND USE OF ABSORPTION SPECTRA

6-5 We may immediately suggest an explanation for the color of Ti_2O_3. The absorption spectrum of the solid indicates a transition occurring in the vicinity of 241 kJ/mol. A study of the emission spectrum of isolated Ti^{3+} ions yields no electronic change which can be identified with this number. If we associate this energy with an electron jump in the presence of the crystal field from the lower *d* orbitals to one of the higher ones (Fig. 6-6), we can see that the splitting parameter Δ for Ti^{3+} must have a value of 241 kJ/mol. Later we shall show that the Δ values obtained in this way are consistent with other experimental values, which indicates that the transition identification is correct.

Absorption spectra thus provide powerful tools for obtaining the Δ parameters and, through the energy-balance relationship, the crystal field stabilization energy. For Ti^{3+}, for example, if Δ is 241 kJ/mol, then the *CFSE*, being $\frac{2}{5}\Delta$, is 96 kJ/mol. Table 6-1 contains the Δ values gleaned from spectral studies of many of the doubly positive, hexahydrated transition-metal ions in solution. In these systems the perturbing elements are the lone-pair electrons on the water molecules, arranged octahedrally around the metal ion. We shall use the numbers of Table 6-1 in Sec. 6-9 to explain the heats of hydration of these molecules.

TABLE 6-1
Some Properties of Doubly Positive Transition-metal Ions

Ion	Configuration	Theoretical μ for isolated ion, BM	*CFSE* in weak octahedral field	*CFSE* in strong octahedral field†	Spectroscopic Δ values for hexahydrated ions, kJ/mol
Ca^{2+}	$(Ar\ core)^{18}\ 3d^0$	0	0	0	
Sc^{2+}	$(Ar\ core)^{18}\ 3d^1$	1.73	$\frac{2}{5}\Delta$	$\frac{2}{5}\Delta$	
Ti^{2+}	$(Ar\ core)^{18}\ 3d^2$	2.83	$\frac{4}{5}\Delta$	$\frac{4}{5}\Delta$	
V^{2+}	$(Ar\ core)^{18}\ 3d^3$	3.87	$\frac{6}{5}\Delta$	$\frac{6}{5}\Delta$	150
Cr^{2+}	$(Ar\ core)^{18}\ 3d^4$	4.90	$\frac{3}{5}\Delta$	$\frac{8}{5}\Delta - W$	166
Mn^{2+}	$(Ar\ core)^{18}\ 3d^5$	5.92	0	$\frac{10}{5}\Delta - W'$	93
Fe^{2+}	$(Ar\ core)^{18}\ 3d^6$	4.90	$\frac{2}{5}\Delta$	$\frac{12}{5}\Delta - W''$	124
Co^{2+}	$(Ar\ core)^{18}\ 3d^7$	3.87	$\frac{4}{5}\Delta$	$\frac{9}{5}\Delta - W'''$	111
Ni^{2+}	$(Ar\ core)^{18}\ 3d^8$	2.83	$\frac{6}{5}\Delta$	$\frac{6}{5}\Delta$	101
Cu^{2+}	$(Ar\ core)^{18}\ 3d^9$	1.73	$\frac{3}{5}\Delta$	$\frac{3}{5}\Delta$	150
Zn^{2+}	$(Ar\ core)^{18}\ 3d^{10}$	0	0	0	

†Strong fields and the significance of *W* are discussed in Sec. 6-7.

(a)

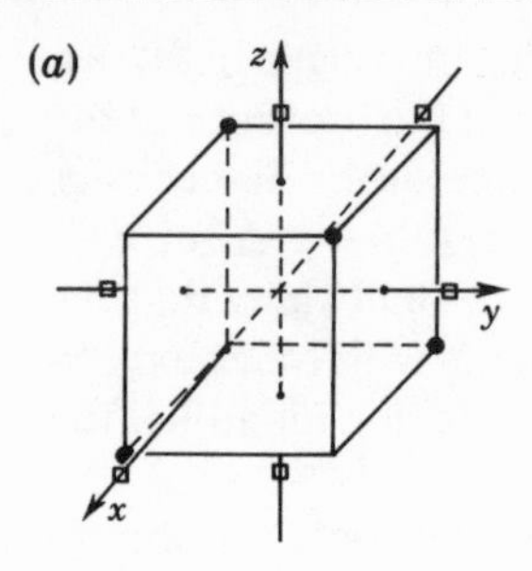

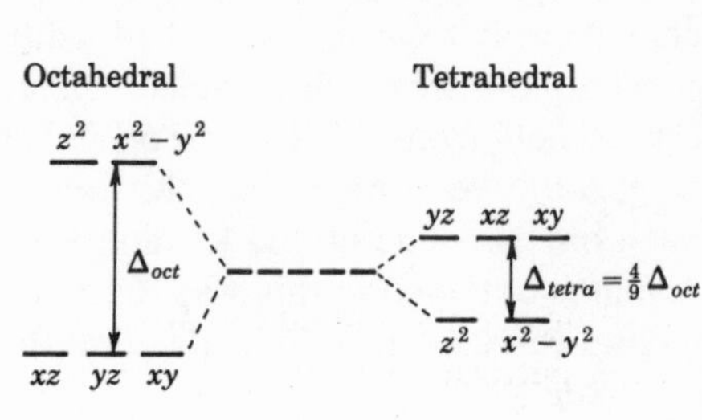

(b)

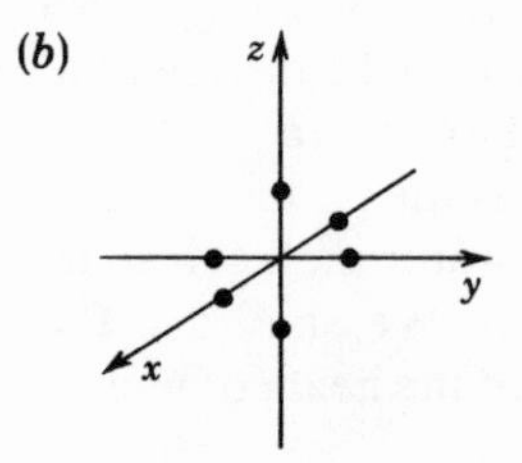

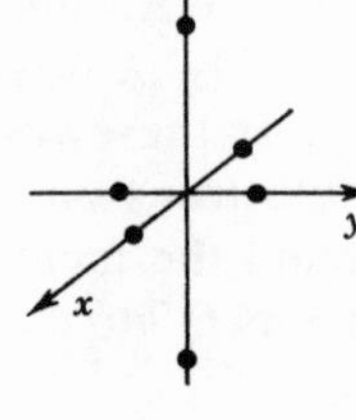

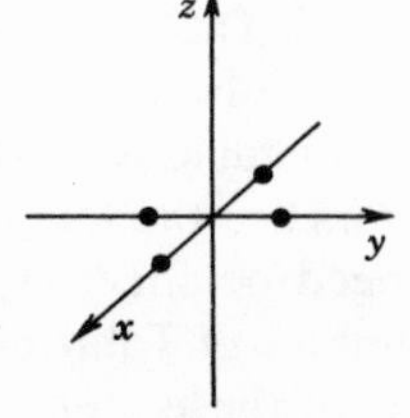

Octahedral Tetragonal

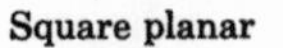

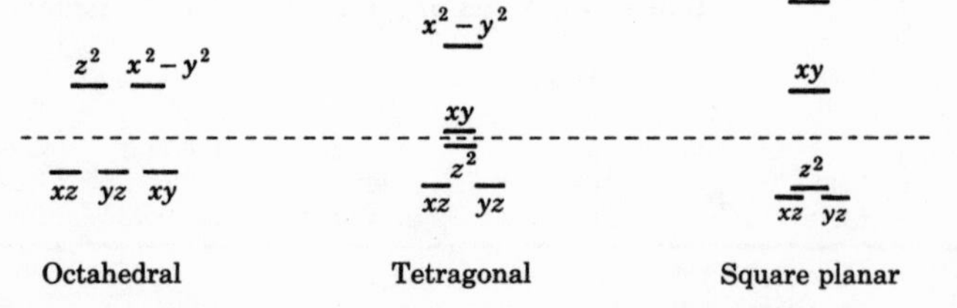

Octahedral Tetragonal Square planar

FIGURE 6-8
Fields of symmetry other than octahedral. (*a*) A tetrahedral array of ligands (black circles) with respect to an octahedral array (squares) and the associated splitting diagrams; (*b*) octahedral, tetragonal, and square planar arrays and their splitting diagrams.

NONOCTAHEDRAL SYSTEMS

6-6 Certainly not all transition-metal ions are surrounded by anions or molecules in octahedral symmetry in their compounds. Other common geometries are square planar, tetrahedral, trigonal, tetragonal, and linear, and each arrangement of perturbing agents (called ligands) produces its own distinctive splitting pattern. We illustrate some of these in Fig. 6-8.

Tetrahedral symmetry is closely related to octahedral symmetry. Figure 6-8*a* shows how one may construct a tetrahedron by placing ligands on four appropriate corners of a cube. In this diagram the x and y axes emerge from the sides of the cube and the z axis from the center of the top face. The tiny black squares show the relative positions of ligands in octahedral symmetry, with metal-ligand distances the same as those in the tetrahedron. Note that the tetrahedral ligands avoid the density lobes of the d_{z^2} and $d_{x^2-y^2}$ orbitals and are more likely to perturb the other three d orbitals. It is not very surprising then to learn that the splitting pattern is inverted when octahedral is replaced by tetrahedral symmetry. Rigorous mathematical treatments show that the splitting associated with tetrahedral symmetry is smaller than that of octahedral; this is logical since the number of ligands is smaller. For equivalent ligand charges and distances.

$$\Delta_{tetra} = \tfrac{4}{9}\Delta_{oct}$$

Figure 6-8*b* illustrates the changes which occur in splitting pattern when, starting with a perfect octahedron, one stretches the ligand-metal distances along the z axis, to form a system of *tetragonal* symmetry, and then removes them completely, leaving the metal ion in a *square planar* field.

SYSTEMS WITH MORE THAN ONE *d* ELECTRON; STRONG AND WEAK CRYSTAL FIELDS

6-7 The general features of all the splitting diagrams discussed so far are the same when the metal ion possesses more than one d electron. However, we must now consider the existence of interactions among the d electrons themselves and the effect of this on the crystal field stabilization energy.

We consider first the case of an ion with two d electrons in a site of octahedral symmetry. We know now what will happen to the first d electron. The second will simply join the first in the lower energy level, as shown in Fig. 6-9, occupying a separate orbital in accordance with Hund's rule, and the system will attain a total $CFSE$ of $2(\frac{2}{5}\Delta)$, or $\frac{4}{5}\Delta$. Since the spin configuration is the same as that of the isolated ion, there will be no significant energy effects due to changes in d electron interactions. A third d electron should behave similarly and increase the crystal field stabilization energy by $\frac{2}{5}\Delta$ more.

A fourth d electron, however, is faced with a difficult decision: it may go down to the lower energy level and pair with one of the repulsive electrons already there, or it may go to the higher, less stable energy level and remain unpaired. If it goes down, the stability of the system increases by $\frac{2}{5}\Delta$ *minus* the interaction energy or work W which must be expended when the electrons pair. If the electron goes to the higher level, the very stable isolated-ion spin configuration is retained (nothing is lost due to electron interaction), but $\frac{3}{5}\Delta$ is subtracted from the $CFSE$ of the system. The decision of the fourth electron (and the fifth and the sixth) is determined by the magnitude of the splitting parameter Δ (the so-called crystal field *strength*) relative to the work of pairing. Two extreme cases can be distinguished, and many, but not all, transition-metal systems fall into one of them. In general, if Δ is very large (if the field is strong), electrons will tend to pair in the lower level at the expense of the associated repulsion energies and form what are called *low-spin* configurations. If the field is very weak (if Δ is small), then the d electrons will retain their maximum-spin configuration, the *high-spin* case, and sacrifice $CFSE$ in so doing. More specifically, for the d^4 case, if $\frac{3}{5}\Delta$ is greater than $\frac{8}{5}\Delta - W$, then the high-spin configuration will be favored. If the reverse of the inequality is true, then the low-spin configuration will be favored.

Strong- and weak-field distributions for all d^n configurations are shown in Fig. 6-9. Note that for strong fields the electrons always pair in the lower level first, filling it before entering the

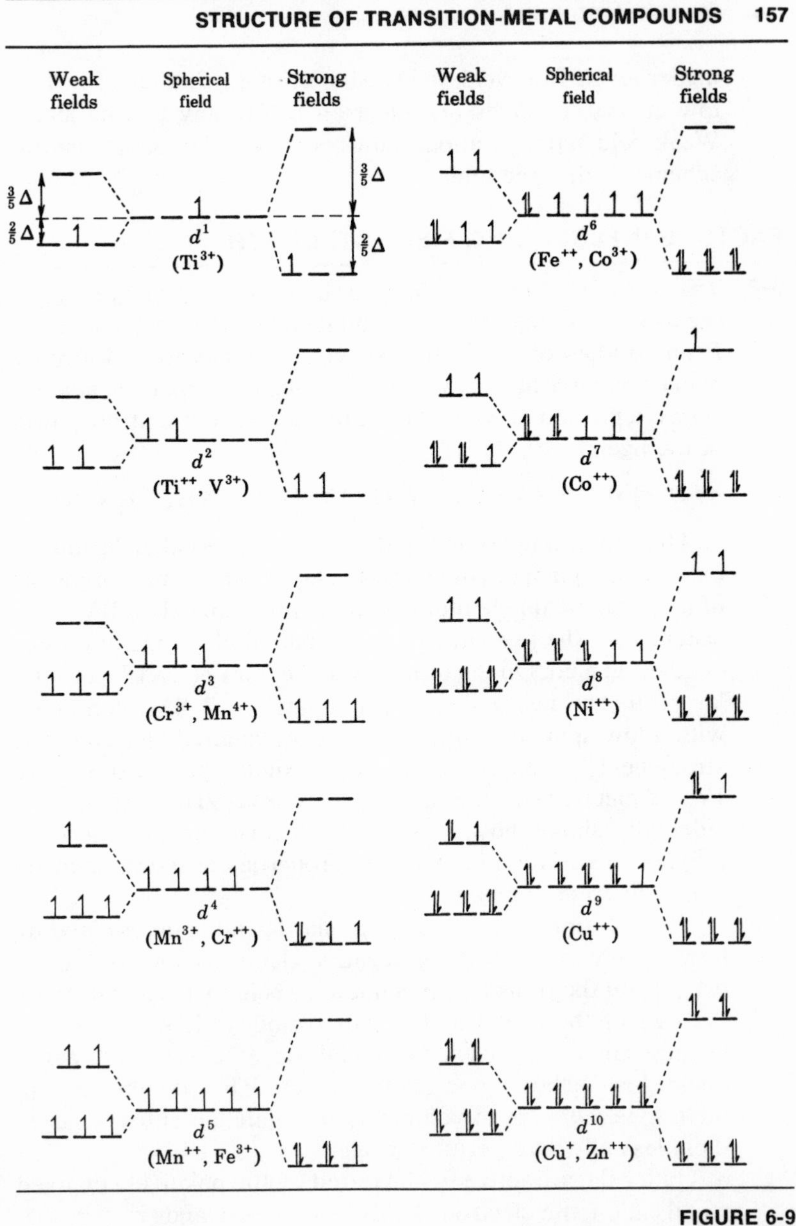

FIGURE 6-9
d-electron configurations for weak and strong fields of octahedral symmetry.

higher level. For weak fields each orbital, whether of high or low energy, receives one electron before any pairing starts. Weak-field configurations always have the same pairing scheme as the free ions.

FACTORS INFLUENCING FIELD STRENGTH

6-8 The strength of the field around an ion depends largely upon the ligand, its size and shape and distance from the metal ion. From studies of absorption spectra it is possible to arrange many common ligands in what is called the *spectrochemical series*, a portion of which, indicating increasing splitting power of the ligand, is:

$$I^- < Br^- < Cl^- < F^- < H_2O \cong O^{2-} < NH_3 \lll CN^-$$

Thus the complex ion FeF_6^{3-} with the weaker ligand F^- exists in a high-spin configuration, as shown by measurements of its magnetic dipole moment. μ is approximately 6 BM, consistent with the presence of five unpaired electrons in a weak octahedral field. On the other hand, the ion $Fe(CN)_6^{3-}$, of similar geometry, has a μ of approximately 1.7 BM, consistent with a low-spin configuration and one unpaired electron. The strong field of the CN^- ion causes maximum pairing of the five Fe^{3+} *d* electrons in the lower energy level. This example incidentally shows how crystal field theory explains the once perplexing variation in magnetic properties of the same metal ion in different compounds.

The splitting power of a particular ligand is determined by how closely the perturbing negative electrons approach the *d* orbitals on the transition-metal ion. This in turn depends upon the size of the ligand and its polarizability. Large, unwieldy ligands cannot approach the metal ion as closely as smaller ones; they "elbow" one another away. This, combined with their lower charge-to-volume ratio (ionic potential), makes them less effective perturbing agents.

Under the attractive field exerted by the positively charged metal cation, the electron clouds on certain ligands may be distorted toward the metal, thus increasing the perturbation of the *d* electrons. Highly electronegative (electron-greedy) ligands resist distortion (polarization) and would be expected to be weaker in splitting power.

Despite these rationalizations the power of a ligand cannot always be predicted satisfactorily. We would expect, on the basis of size and ionic potential, that I^- relative to F^- should be the weaker liquid. However, since F is more electronegative, we would expect that I^- should be more polarizable and consequently the stronger ligand. Experimental results indicate that I^- is weaker (at least in interactions with transition metals), and the main conclusion we can draw from this is that here, as in all of chemistry, the court of last appeal is the experiment.

EXPLANATION OF THE STRANGE DOUBLE-HUMPED CURVES

6-9 One of the biggest accomplishments of crystal field theory is the satisfactory explanation of the twin peaks observed when many physical properties of transition metals are plotted versus atomic number. Figure 6-10 illustrates such a variation in the heats of hydration of the doubly positive ions. The heat

FIGURE 6-10
Experimental and "corrected" heats of hydration of doubly positive transition-metal ions.

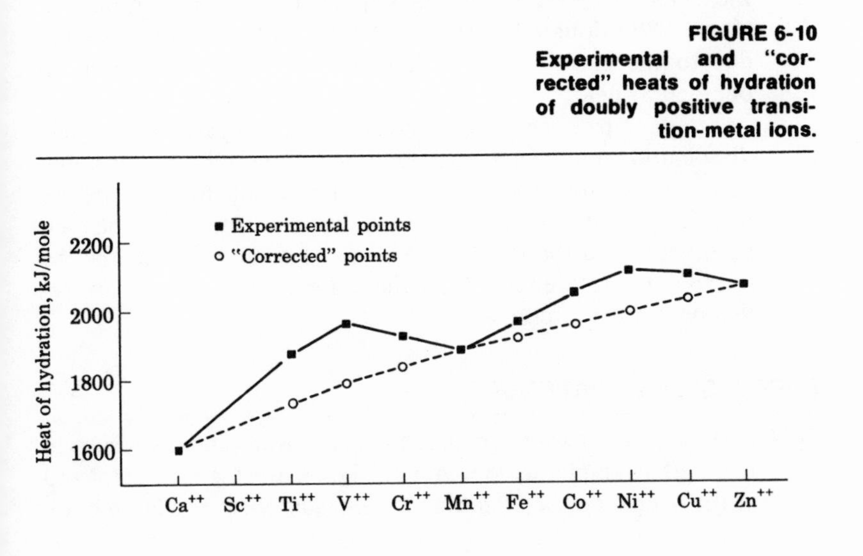

of hydration, E_h, is the total increase in stability experienced by the isolated ion when it becomes surrounded by about six water molecules in solution, i.e.,

$$Tr^{2+} + 6H_2O \rightarrow Tr(H_2O)_6^{2+} + E_h$$

E_h is the energy which must be supplied to the complex ion to reverse the hydration process (in theory). If we were unaware of crystal field effects we would expect this energy to increase slowly and smoothly from Ca^{2+} to Zn^{2+} because of the ligands' attraction for the increasing effective charge on the metal ion. Although all ions in the series are formally doubly positive species, their *effective* charges are not +2. From Ca^{2+} to Zn^{2+} the nuclear charges increase regularly; yet each increase is only partially canceled by the increase in the number of *d* electrons externally. The *d* electrons are relatively poor shielders of one another from the nucleus.

The experimental curve of hydration energies, however, shows two distinct peaks, with a minimum occurring at Mn^{2+}. If we subtract from this curve the crystal field stabilization energies of doubly positive ions in a weak field (using stabilization-energy expressions and Δ values of Table 6-1), the result is the relatively smooth curve expected from our arguments above. The double-humped character of the curve is simply due to the extra stability attained by the *d* electrons in an octahedral field.

Similar curves showing trends in ionic radii, lattice energies, dissociation energies of gaseous molecules, stability constants of complex ions, and even reaction rates on the surfaces of transition-metal compounds have been beautifully explained by crystal field theory. Descriptions of some of these may be found in the references at the chapter end, while others are developed in the exercise section.

EFFECTS OF COVALENCE

6-10 Although so far we have described transition-metal systems in terms of a completely ionic model, i.e., with electrons localized at the transition metal and on the ligands, there is a large body of experimental evidence indicating that delocalization and consequent partial covalence do occur.

Although the ionic model explains the main features of the absorption spectra of many transition-metal compounds, experimental and theoretical results do not often agree exactly. In many compounds the actual metal-ligand distances are much smaller than the sum of ionic radii. Lattice energies computed for many of the salts and oxides of transition metals (even with elaborate corrections) are quite different from experimental values. All these differences may be attributed to the effect of covalence, i.e., delocalization of electron clouds between metal and ligands. The idea of partial covalence in these compounds is not surprising, for by now we know that few systems belong to the extreme classifications purely ionic and purely covalent.

The molecular-orbital theory of transition-metal compounds takes into account the mixing of ligand and metal electron clouds and is consequently favored over the ionic model by many chemists. Molecular-orbital *calculations*, however, have thus far proved no more successful in *quantitative* explanation of properties of transition-metal compounds than those based on the ionic model.

To build molecular orbitals for an octahedral complex, generally one assumes that six σ bonds may form by overlap of ligand orbitals with metal orbitals directed along the x, y, and z axes: $4s$, $4p_x$, $4p_y$, $4p_z$, $3d_{z^2}$, $3d_{x^2-y^2}$. The $3d_{xy}$, $3d_{xz}$, and $3d_{yz}$ orbitals may be assumed *nonbonding* or in some systems may be used to form π *MO*s by sideways overlap with ligand p orbitals.

Figure 6-11 illustrates a molecular-orbital energy diagram for an octahedral complex where for simplicity only σ *MO*s are considered. The six ligand and six metal orbitals smear together to form six partially degenerate *bonding* molecular orbitals (whose levels are labeled σ_1, σ_2, and σ_3) and six *antibonding MO*s (with levels labeled σ_1^*, σ_2^*, and σ_3^*). Electrons from both metal and ligands occupy these *MO*s. In the $Ti(H_2O)_6^{3+}$ complex ion, for example, one lone pair on each of the six water molecules contributes to the σ bond formation,

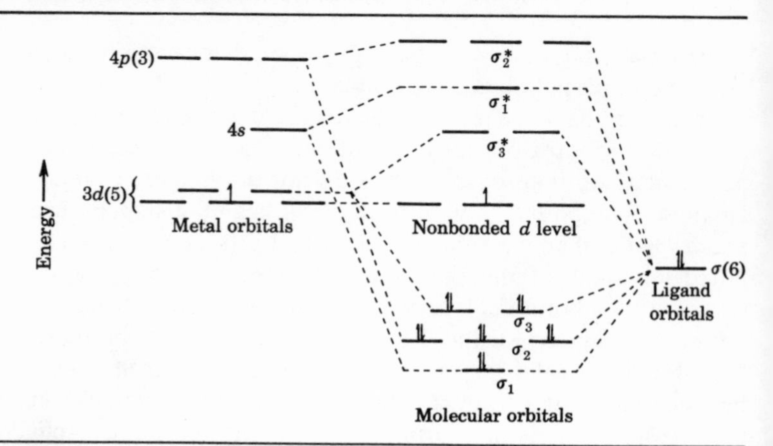

FIGURE 6-11
Molecular-orbital energy diagram for the octahedral complex ion $Ti(H_2O)_6^{3+}$. The numbers in parentheses represent degeneracies. Dotted lines indicate the metal and ligand orbitals mixed to build the *MO*s.

filling the lowest six molecular orbitals in Fig. 6-11. The nonbonding d level contains the $Ti^{3+}(d^1)$ electron. The absorption spectrum of $Ti(H_2O)_6^{3+}$ is now attributed to a transition from this triply degenerate ground state to the doubly degenerate σ_3^* level, composed partly of $d_{x^2-y^2}$ and d_{z^2} orbitals, an identification differing from that of the ionic model only in that here not all the d orbitals involved are "pure."

The amount of ligand-metal electron mixing necessarily varies from compound to compound. We may formulate some general rules about the degree of covalence to be expected, basing our arguments on the concept of ionic potential (Chap. 5). For similar compounds, FeO, MnO, CoO, NiO, we expect covalence to increase with decrease in metal-ion size. For series like TiO_2, V_2O_5, CrO_3, Mn_2O_7, we expect increasing covalence with increase in oxidation number of the metal, and indeed, TiO_2 is a crystalline solid while Mn_2O_7 is a green gas made up of covalent molecules.

Strong-field ligands (within the ionic model) such as CN^- in the complex ion $Fe(CN)_6^{3-}$ overlap strongly with metal orbitals. As a result, the energies of the molecular orbitals involving the $d_{x^2-y^2}$ and d_{z^2} orbitals (σ_3 and σ_3^*) become respectively highly stable and unstable with respect to the nonbonding d level, and the Fe^{3+} d electrons pair in a *low-spin* configuration in the nonbonding d level after 12 CN^- lone-pair electrons have filled σ_1, σ_2, and σ_3 levels (Fig. 6-12*a*). In contrast, in the ion FeF_6^{3-} the F^- electron pairs are not easily distorted toward the metal and overlap is smaller. The *MO*s likely to form do not differ appreciably in energy from the isolated orbitals; i.e., there is less stabilization due to mixing of electron clouds (Fig. 6-12*b*), and the outer electrons occupy the nonbonding d level

FIGURE 6-12 Molecular-orbital energy diagrams for complex ions of Fe^{3+}. (*a*) Covalent, low-spin $Fe(CN)_6^{3-}$, for which *MO* energies are considerably different from those of the separated ions; (*b*) the "ionic," high-spin FeF_6^{3-}, for which *MO* energies differ only slightly from energies of separated ions.

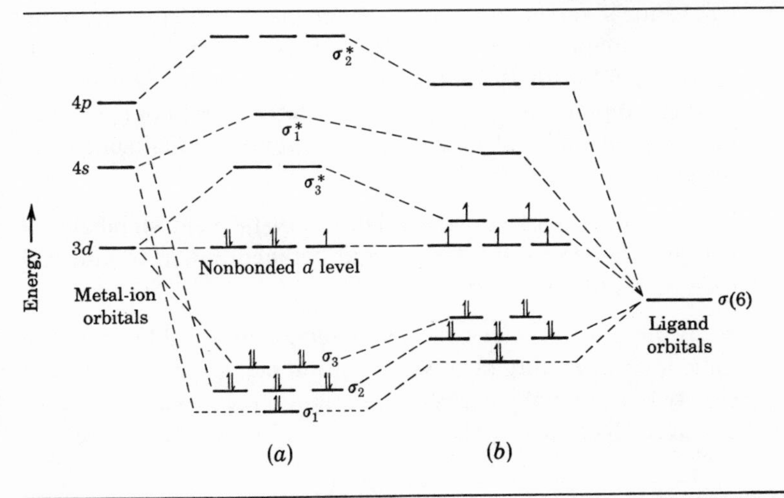

and the nearby σ_3^* level in a *high-spin* configuration. Roughly correlating the two models, we may say that strong-field complexes (within the ionic model) correspond to highly covalent systems (within the *MO* treatment), while the weak-field complexes are only slightly covalent.

Perhaps more quantitative things may be said about the degree of covalence in the compounds when many rigorous calculations, yielding accurate descriptions of electron density distributions, have been performed.

EXERCISES

1 The $Ti(H_2O)_6^{3+}$ complex ion shows a strong absorption at 490 nm. Predict whether $Ti(NH_3)_6^{3+}$ would absorb light of a higher or lower wavelength.

2 Consider each of the species Cr^0, Cr^{2+}, and Cr^{3+} in an octahedral field. (*a*) Use a suitable diagram to represent the occupation of energy levels for weak and strong fields around the metal or its ions. (*b*) Explain how magnetic properties may be used to distinguish the strong and weak extremes for each case.

3 The complex ion $Ni(CN)_4^{2-}$ exists in a square planar configuration. Keeping in mind the position of CN^- in the spectrochemical series, compute in Bohr magnetons the magnetic dipole moment expected for the salt $Na_2[Ni(CN)_4] \cdot 3H_2O$. (See Fig. 6-8*b* for the square planar splitting diagram.) Compute the dipole moment assuming a tetrahedral configuration of CN^- ions. Would magnetic measurements distinguish between these two forms?

4 Calculate in kJ/mol the relative crystal field stabilization energy attained by the Fe^{2+} ion in octahedral and tetrahedral oxide-ion environments.

5 If there were no crystal field effects, one would expect the ionic radii of doubly positive transition-metal ions to decrease slowly but gradually across the periodic table. Why? Actually a "curve" like the one shown in Fig. 6-13 is observed. Attempt

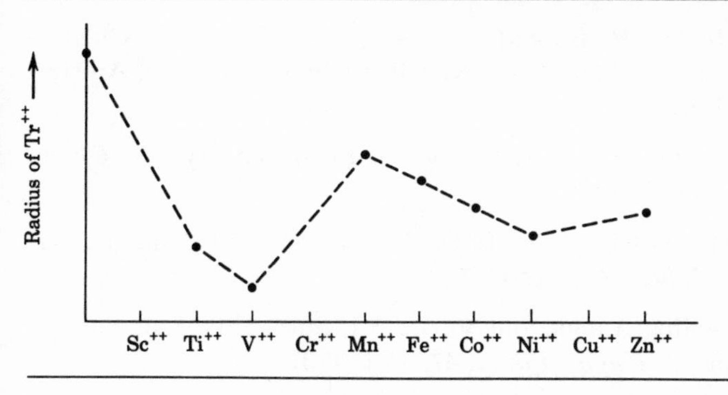

FIGURE 6-13
Variation in metal-oxygen distances in oxides.

an explanation, assuming a perfect octahedral weak field of oxide ions around the metal ion.

REFERENCES FOR FURTHER READING

1 Orgel, L. E., "An Introduction to Transition Metal Chemistry: Ligand Field Theory," 2d ed., Methuen & Co., Ltd., London, 1966.

2 Halliday, D., and R. Resnick, "Physics for Students of Science and Engineering," 2d ed., part II, chap. 37: Magnetic Properties of Matter, John Wiley & Sons, Inc., New York, 1962.

3 Bitter, F., "Magnets: The Education of a Physicist," Doubleday & Company, Inc., Garden City, N.Y., 1959.

4 Pearson, R. G., Crystal Field Explains Inorganic Behavior, *Chem. Eng. News*, **29,** 72 (1959).

5 Sutton, L., Some Recent Developments in the Theory of Bonding in Complex Compounds of the Transition Metals, *J. Chem. Educ.*, **37,** 498 (1960).

6 O'Reilly, D., Electronic Structure of Metal Oxides, *J. Chem. Educ.*, **38,** 312 (1961).

7 Heslop, R. B., and P. L. Robinson, "Inorganic Chemistry," 3d ed., chap. 8, Elsevier Publishing Company, Amsterdam, 1967.

8 Cotton, F. Albert, Ligand Field Theory, *J. Chem. Educ.*, **41,** 466 (1964).

9 Zuckerman, J. J., Crystal Field Splitting Diagrams, *J. Chem. Educ.*, **42,** 315 (1965).

10 Pauling, Linus, and Vance McClure, Five Equivalent *d* Orbitals, *J. Chem. Educ.*, **47,** 15 (1970).

11 Doyle, W. P., Principles of Ligand Field Theory, *Chemistry*, **39,** 14, January (1966).

PERIODIC TABLE
APPENDIX A

Period	Group Ia	IIa	IIIa	IVa	Va	VIa	VIIa	VIII			Ib	IIb	IIIb	IVb	Vb	VIb	VIIb	0
1	1 **H**																1 **H**	2 **He**
2	3 **Li**	4 **Be**											5 **B**	6 **C**	7 **N**	8 **O**	9 **F**	10 **Ne**
3	11 **Na**	12 **Mg**					Transition series						13 **Al**	14 **Si**	15 **P**	16 **S**	17 **Cl**	18 **Ar**
4	19 **K**	20 **Ca**	21 **Sc**	22 **Ti**	23 **V**	24 **Cr**	25 **Mn**	26 **Fe**	27 **Co**	28 **Ni**	29 **Cu**	30 **Zn**	31 **Ga**	32 **Ge**	33 **As**	34 **Se**	35 **Br**	36 **Kr**
5	37 **Rb**	38 **Sr**	39 **Y**	40 **Zr**	41 **Nb**	42 **Mo**	43 **Tc**	44 **Ru**	45 **Rh**	46 **Pd**	47 **Ag**	48 **Cd**	49 **In**	50 **Sn**	51 **Sb**	52 **Te**	53 **I**	54 **Xe**
6	55 **Cs**	56 **Ba**	57* **La**	72 **Hf**	73 **Ta**	74 **W**	75 **Re**	76 **Os**	77 **Ir**	78 **Pt**	79 **Au**	80 **Hg**	81 **Tl**	82 **Pb**	83 **Bi**	84 **Po**	85 **At**	86 **Rn**
7	87 **Fr**	88 **Ra**	89** **Ac**															

* Lanthanide series	58 **Ce**	59 **Pr**	60 **Nd**	61 **Pm**	62 **Sm**	63 **Eu**	64 **Gd**	65 **Tb**	66 **Dy**	67 **Ho**	68 **Er**	69 **Tm**	70 **Yb**	71 **Lu**
** Actinide series	90 **Th**	91 **Pa**	92 **U**	93 **Np**	94 **Pu**	95 **Am**	96 **Cm**	97 **Bk**	98 **Cf**	99 **Es**	100 **Fm**	101 **Md**	102 **No**	103 **Lw**

Periodic table of the elements

ANSWERS OR HINTS TO SOLUTION OF SELECTED EXERCISES

APPENDIX B

CHAPTER 1

1 violet, red

2 wave

3 corpuscular

4 10^3 m/s

5 *d*

6 0.485 nm

7 4.97×10^{-19} J

8 longer, *B*, greater, λ_2

CHAPTER 2

1 $-K/4, -K/25, -19.6 \times 10^{-18}$ J, case *a*, infinite

2 16/15**R**

3 *C*, *C*

4 ultraviolet, 164.1 nm

CHAPTER 3

1 2*d* and 3*f* are not possible.

2 18, 2, 14, 50

6 *Hint:* consider the increase in nuclear charge and also the effect of Hund's rule.

9 (*c*) Na > Al > P > Ar
Sr > Ge > Si > Ar
Sb > Se > Cl > Ne
Te > Se > S > O

CHAPTER 4

1 With no unpaired electrons there must be a double bond between nitrogen atoms. Cis and trans forms would not easily interchange. Cis would have a dipole moment.

3 NH_4^+, sp^3; H_2Se, none; NO_2^+, sp; SO_3, sp^2.

5 Central carbon, *sp;* terminal carbons, sp^2; the planes containing the CH_2 groups must be mutually perpendicular to permit formation of two separate π bonds.

7 F_2

8 NO^+ is isoelectronic with N_2, with $BO = 3$, and accordingly has the strongest bond and the shortest bond distance.

11 Both NO^+ and PS^+ have bond orders of 3. As a molecule of third row atoms, PS^+ has thinner electron glue. Consequently NO^+ is most stable. $ArCl^+$ with a bond order of 1 is least stable.

12 *Hint:* NO_2^+ is isoelectronic with CO_2. Compare NO_2^- with SO_2.

13 The XeF_4 portion of the molecule may be as illustrated in Fig. 4-30*c*, with Xe lone pairs above and below the XeF_4 plane. The oxygen atom in a non-ground-state $2s^22p_x^22p_y^22p_z^0$ configuration approaches XeF_4 and overlaps its *empty* $2p_z$ orbital with a lone pair on Xe. The pair is shared between O and Xe in a normal covalent bond, sometimes called a *dative* bond, since one atom donates both of the electrons for the bond.

16 The sulfur atom, with configuration $3s^23p^4$, has six valence electrons. Each fluorine atom contributes one electron to the bond region, making a total of 10 electrons or five electron pairs around sulfur. Four of these must be bond pairs; consequently there is one lone pair. The five dsp^3 hybrids of Fig. 4-29*a* can be used to describe a situation where an atom is surrounded by five pairs. Possible geometries are three bonds in the plane, one bond above and a lone pair below, *or* two bonds along the axis, two more bonds in the plane and one lone pair in the plane. The experimental geometry is close to the latter.

19 The BCl_3 molecule must be planar and symmetrical with three bonds separated by 120° so that bond dipole moments cancel. The boron atom, with $2s^2 2p^1$ configuration, must be in a trigonal sp^2 hybrid state with one electron in each hybrid (Fig. 4-16). Sigma bonds are formed through overlap with the unpaired $3p$ electrons on the chlorine atoms. The $2p_z$ orbital on boron, perpendicular to the σ framework, is *empty*, and the molecule is called "electron-deficient." (Some delocalized π bonding is possible through overlap of $3p_z$ orbitals on Cl atoms with the empty $2p_z$ on B, but because of size difference this should not be extensive.)

20 *Hint:* borazole is isoelectronic with benzene, C_6H_6. Also consider the fact that boron is less electronegative than nitrogen.

21 In diamond all carbon atoms are in sp^3 hybrid states and are connected to one another by strong σ bonds. In graphite all carbon atoms are in sp^2 hybrid states, are connected to one another by σ bonds, and are furthermore held together in a layer by a π structure extending over all carbon atoms. (Consider the description given for the benzene molecule, remove H atoms, and extend bonds in all directions to other carbon atoms.) In graphite the layers are held together by van der Waals forces. (These are discussed in Chap. 5.)

CHAPTER 5

1 $E_{ic} = 5.82E_{ip}$, $U_{ic} = 1.46NE_{ip} = 718$ kJ/mol.

2 Assuming that $d = 0.207$ nm (the sum of the ionic radii) we find that the stabilization energy is 1804 kJ/mol.

3 Note that E_{rep} is U_L/n or 94 kJ/mol. The corrected lattice energy is thus 766 kJ/mol.

4 (*a*) On the basis of ionic potential $BiCl_3$ is most ionic, $BeCl_2$ least ionic.

5 4 or 6; 4 or 6, possibly 8; 4 or 6; 4; 8 or 12.

8 Decreasing H-bonding: CH_3OH, CH_3NH_2, CH_3F, CH_4.

9 A is unlikely because of the small size of the proton; B is possible.

11 *Hint:* What sort of forces hold I_2 molecules together in solid I_2?

12 Molecules are held together within a sheet by hydrogen bonds. Van der Waals forces hold the layers together.

13 The interionic distance in LiF should be smaller than that in LiH. Accordingly, the lattice energy of LiF should be larger than that of LiH. Ions in LiF should be more tightly bound and consequently LiF should have a higher melting point. Does it?

14 (*a*) $O^{2-} > F^-$ and $S^{2-} > Cl^-$ (rule 1); $S^{2-} > O^{2-}$ (rule 2); consequently S^{2-} is largest.

15 Cs, thinnest electron glue

16 poorer; the alloy is stronger since C atoms inhibit slip.

19 S atoms are closer in size to P atoms and would most likely substitute for P atoms in the lattice. With one extra electron over P atoms, S atoms would most likely cause *n*-type semiconduction.

20 $Si_3O_{10}^{8-}$, $Si_4O_{12}^{8-}$

21 *a*, *b*, *e*

CHAPTER 6

1 $Ti(NH_3)_6^{3+}$ would absorb light of a lower wavelength than 490 nm, corresponding to an electron transition of higher energy. NH_3 is a stronger ligand than H_2O.

3 For square planar geometry, $\mu = 0$; for a tetrahedral array of CN^- ions, $\mu = 2.83$ BM.

4 From the spectrochemical series Δ for O^{2-} is approximately equal to that of water. In octahedral fields Δ for Fe^{2+} is 124 kJ/mol and the *CFSE*, $\frac{2}{5}\Delta$, is 50 kJ/mol. The tetrahedral Δ is $\frac{4}{9}\Delta_{oct}$ or 55 kJ/mol, and the *CFSE* is three-fifths of this or 33 kJ/mol.

5 *Hint:* Compare the relative effectiveness of electrons in d_{xy}, d_{xz}, and d_{yz} orbitals with those in d_{z^2} or $d_{x^2-y^2}$ orbitals in shielding ligands from the nuclear charge on the metal.

INDEX